BRAZILIAN STONE METEORITES

BRAZILIAN STONE METEORITES

With a brief general introduction on the significance, classification, mineralogy, bulk composition, and recognition of stone meteorites.

Celso B. Gomes
and
Klaus Keil

UNIVERSITY OF NEW MEXICO PRESS
Albuquerque

Library of Congress Cataloging in Publication Data

Gomes, Celso B.
Brazilian stone meteorites.

Bibliography: p.
Includes index.
1. Chondrites (Meteorites) 2. Meteorites—Brazil.
I. Keil, Klaus, joint author. II. Title.
QB755.K44 523.5′1′0981 80-5333
ISBN 0-8263-0543-1

Manufactured in the United States of America.
Library of Congress Catalog Card Number 80-5333
International Standard Book Number 0-8263-0543-1
First Edition.

CONTENTS

1. PREFACE

Our interest in Brazilian stone meteorites began in 1974 when one of us (K.K.) spent part of his sabbatical leave at the Instituto de Geociências, Universidade de São Paulo, São Paulo, Brazil. Two facts regarding Brazilian stone meteorites became immediately apparent: first, many had only been poorly described and, in many instances, even the most basic data were either not available or the analytical data were so old as to render them doubtful. Secondly, among the Brazilian stone meteorites are some unique ones which, on the basis of limited data existing at the time, appeared particularly significant in terms of their importance for ideas on the origin and evolution of their parent bodies and, indeed, for the early history of our solar system. In this category, for example, were the achondrites Angra dos Reis and Serra de Magé. Considering these two facts, we decided to begin a systematic study of all Brazilian stone meteorites. In the case of common, run-of-the-mill chondrites, we carried out basic mineralogic, petrographic, electron microprobe, and bulk chemical studies in order to provide a data base for these stones. In the case of unique meteorites, such as Angra dos Reis, Serra de Magé, Governador Valadares, and Ibitira, consortia were organized involving studies not only of mineralogy, petrology, and bulk composition, but also of absolute age dating, trace element geochemistry, crystallography, etc. These papers have been published in various journals and are listed in the reference section (9).

The purpose of the present book is several fold. First, we felt a need existed for summarizing all data on Brazilian stone meteorites in one place. This is done in handbook format in chapter 7. We have attempted to cover the literature as completely as possible but decided to delete clearly erroneous or outmoded data, although reference to these is always made. Second, we also felt that a brief introduction into some of the basic aspects of meteoritics in this book would be useful (chapters 2–6). There, we emphasize the scientific significance of meteorites, their classification, mineralogy, bulk composition and recognition. No attempt is made to make this into a textbook on meteorites. Rather, we hope that chapters 2–6 will help serve as a basis for understanding and appreciating the importance of meteorites in general and the data summarized in the chapter on Brazilian stone meteorites in particular. We also hope that this book will serve to stimulate the interest of Brazilian laymen, students, and scientists in meteorites, their study, and recovery.

For details on other aspects of meteoritics, the reader is referred to the books by Cohen (1894, 1903, 1905), Farrington (1915), Perry (1944), Brown et al. (1953), Krinov (1960), Mason (1962, 1971), Heide (1964), Hey (1966), Wood (1968), Ramdohr (1973), Wasson (1974), Buchwald (1975), Nagy (1975), Hutchison et al. (1977), and the review articles by Anders (1963, 1964, 1971a), Wood (1963), Ringwood (1966), Hayes (1967), Mason (1967a, b), Keil (1969a), Dodd (1969), Van Schmus (1969), and Wasson (1972).

2. INTRODUCTION

Meteorites are solid bodies which reach the Earth (or the Moon, Mars, etc.) from interplanetary space and are large enough to survive passage through the Earth's (or Mars', etc.) atmosphere so that pieces can be collected. All evidence indicates that meteorites come from our solar system, although recent work (e.g., Clayton et al., 1973; Clayton, 1978) suggests that some meteorites may contain material that was originally of extra-solar system origin and was mixed with solar system material during the agglomeration of the meteorite parent bodies. When a meteorite is seen to fall and material is collected, the meteorite is referred to as a *fall*. On the other hand, a meteorite *find* is the by-chance recovery of meteoritic material with no record of the fall. A meteoroid is an object larger than an atom or molecule, traveling in space, and a meteor is the light phenomenon associated with the entry of a small meteoroid into the Earth's atmosphere, whereby the object is entirely consumed by ablation. Individual meteorites

Figure 1 – The Bendegó, Bahia, Brazil iron meteorite, weighing approximately 5,360 kg. Note the irregular shape and the extensively carved surface due to ablation upon entry into the earth's atmosphere (with Dr. W.S. Curvello, Museu Nacional, Rio de Janeiro, Brazil).

vary in size from the 60 ton Hoba, Southwest Africa, iron to objects considerably below 0.1 gram in weight. The largest Brazilian meteorite, the Bendegó, Bahia, iron weighs 5.36 tons and is on exhibit in the Museu Nacional, Rio de Janeiro (Fig. 1).

In addition to genuine meteorites, dust of extraterrestrial origin accumulates on Earth but usually goes unnoticed because of the small sizes of individual dust grains. Among these materials, for example, are cosmic dust (micron-to submicron-sized particles small enough to accumulate on the surface of the Earth without undergoing melting during entry); meteoric dust (resulting from the disintegration of meteors); and meteoritic dust (resulting from the ablation and disintegration of meteorites during their entry into the Earth's atmosphere (e.g., Schmidt, 1965). Except for meteoritic dust (e.g., Schmidt and Keil, 1966; Brownlee et al., 1978), study of extraterrestrial dust has been difficult and often frustrating, because of difficulties in its collection and discrimination from terrestrial contamination (e.g., Brownlee et al., 1977a, b, 1978; Millman, 1977; Bibring et al., 1978).

Meteorites are basically divided into three categories, the stones, the stony-irons, and the irons. One major classification criterion is the content of metallic nickel-iron, with the stones having $\lesssim$ 30%, the stony-irons $\gtrsim$ 30 $\lesssim$ 65%, and the irons $\gtrsim$ 95%. The stones are subdivided into the achondrites and chondrites. The achondrites

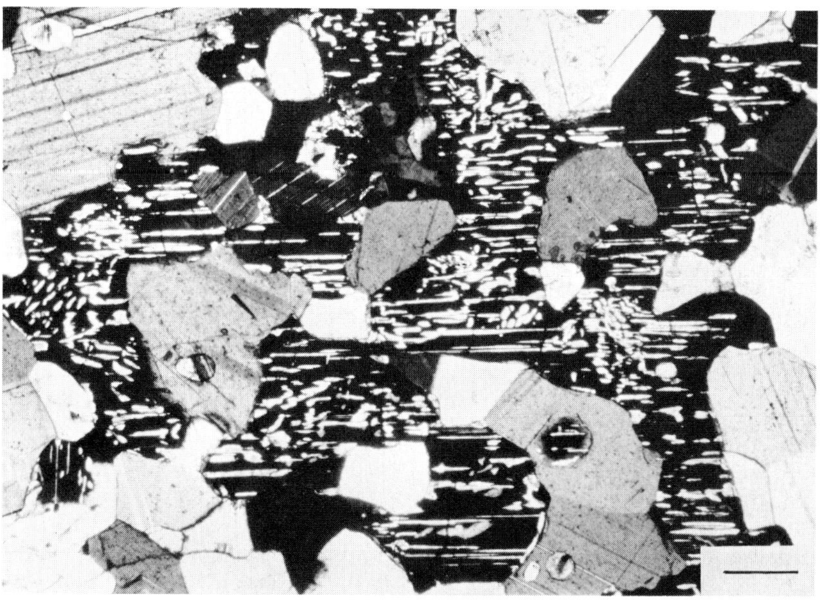

Figure 2 — Photomicrograph of the Serra de Magé, Pernambuco, Brazil monomict pigeonite-plagioclase achondrite (eucrite, non-brecciated), showing an annealed, igneous texture consisting of low-Ca pyroxene with exsolved high-Ca pyroxene; plagioclase is poikilitically enclosed in the pyroxene. Transmitted light, crossed polarizers. Scale bar equals 0.15 mm.

most closely resemble terrestrial ultramafic and mafic rocks (both in composition and texture), but are often brecciated and

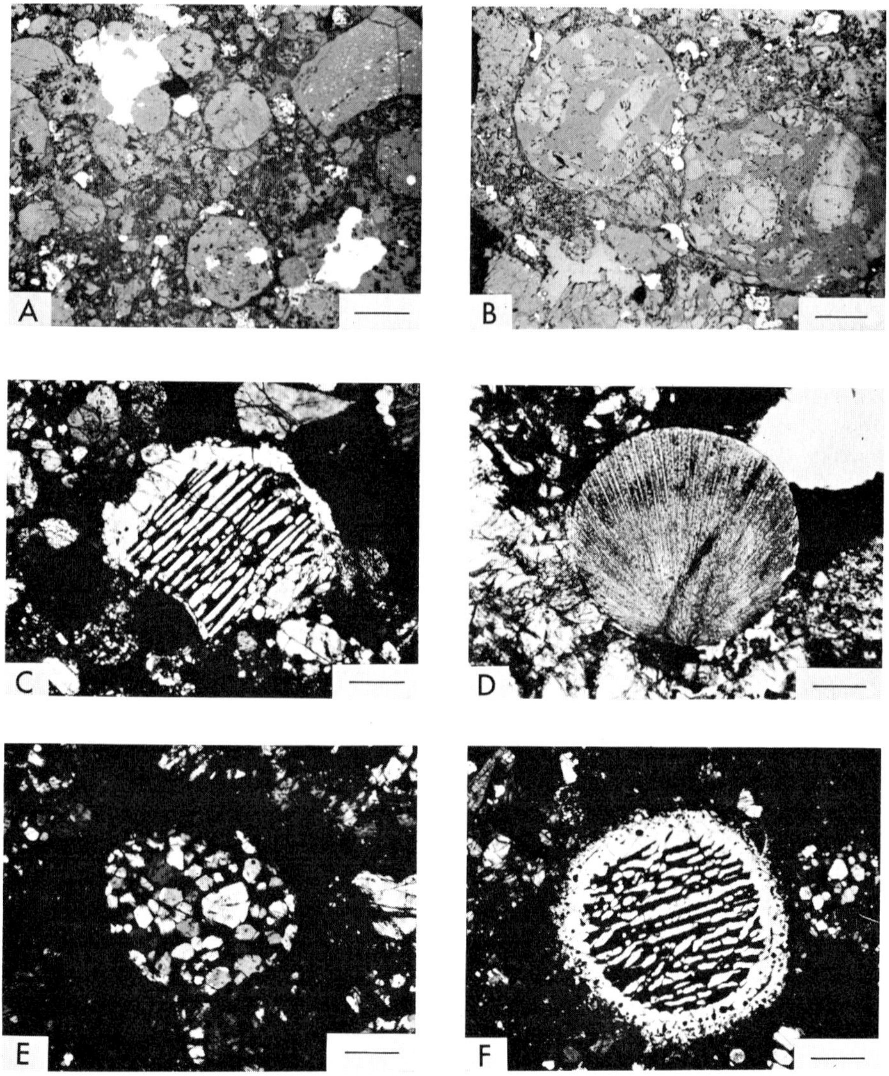

Figure 3 — Photomicrographs of the Conquista, Minas Gerais, Brazil chondrite, showing the well-developed chondritic texture. A) Overview of the texture, showing many chondrules that have sharp, clearly discernible boundaries with the chondrite matrix. White is metallic nickel-iron and troilite, black areas are due to plucking of the friable meteorite in the polishing process. Reflected light, scale bar equals 0.35 mm. B) Porphyritic chondrules containing olivine and pyroxene (light gray) in a glassy to microcrystalline matrix (medium gray). Reflected light, scale bar equals 0.32 mm. C and F) Barred and rimmed olivine chondrules, consisting of bars of olivine (white) in a glassy to microcrystalline matrix (black). Transmitted light, crossed polarizers. Scale bar in C equals 0.09 mm and in F 0.22 mm. D) Excentro-radial pyroxene chondrule. Black is metallic nickel-iron. Transmitted, plane polarized light. Scale bar equals 0.06 mm. E) Granular-porphyritic chondrule, consisting largely of olivine crystals. Transmitted light, crossed nicols. Scale bar equals 0.09 mm.

usually have $\lesssim$ 1% metallic nickel-iron (Fig. 2). The chondrites are also ultramafic rocks but have "chondritic", more or less recrystallized textures, consisting (in most cases, except for carbonaceous chondrites of type I), of millimeter-sized spherules (chondrules) of silicates (olivine, pyroxene, plagioclase, glass), with some metallic nickel-iron and troilites, that are embedded into a matrix of variably-sized silicates and chondrule fragments, fine-grained material, abundant metallic nickel-iron and troilite (Fig. 3) (for details see chapter 4). The stony-irons comprise different sub-types of different mineral contents and textures (the pallasites, siderophyre, lodranite, and mesosiderites; Fig. 4). The iron meteorites consist of varying amounts and textures of two nickel-iron alloys, kamacite (the α-form, $\lesssim$ 7% Ni) and taenite (the γ-form $\gtrsim 7 \lesssim$ 50% Ni) and are divided into the hexahedrites (only kamacite), the octahedrites (varying amounts of kamacite and taenite intergrown along the octahedral crystallographic faces, giving rise to the characteristic "Widmanstätten" structure; Fig. 5), and the ataxites, consisting of fine-grained intergrowths of α and γ metallic nickel-iron.

Meteorites are named after their place of fall or find (nearest city, village, river, landmark, post-office, etc.). Since 1969, Japanese and joint USA-Japanese expeditions have recovered over 1,300 meteorite specimens, representing perhaps as many as 330 separate falls, in relatively small areas on the Antarctic ice near the Yamato Mountains, Mt. Baldr and Allen Hills (e.g., Nagata,

Figure 4 — Handspecimen of the Dora, New Mexico, U.S.A. pallasite, consisting of a continuous network of metallic nickel-iron (white, $\alpha + \gamma$ phases)) and euhedral olivine crystals (gray). Scale bar equals 1.65 cm.

Figure 5 — Polished and etched section of the Bendegó, Bahia, Brazil coarse octahedrite, showing the Widmanstätten structure. Scale bar equals 3.30 cm (Photo by W.S. Curvello).

1975, 1978; Cassidy et al., 1977; Olsen et al., 1978). These discoveries of great numbers of meteorites in such small areas that are largely without landmarks required a new nomenclature for their naming, consisting of the essential landmark (e.g., Allen Hills), followed by a letter indicating the particular expedition and a 5 digit number. The first two digits of this number indicate the December year of the expedition and the next three are numbers given to each specimen as they are collected (note that in this scheme, separate pieces of the same fall receive separate numbers) (Meteorite Nomenclature Committee, Meteoritical Society Meeting, Sudbury, Canada, 1978). There are now 1944 genuine meteorites represented in the world's meteorite collections, of which 765 are observed falls and 1179 are finds (Table 1) (not included in these statistics and table are the meteorites recently recovered in Antarctica, because most have not been thoroughly studied and, thus, the true number of separate falls represented by these finds has not yet been established). Among the falls, stone meteorites (92.9%) dominate over stony-irons (1.4%) and irons (5.6%), whereas among the finds, over 50% are irons and stony-irons. This is the result of the greater resistance of irons and stony-irons to weathering on Earth, as well as of their greater conspicuousness which makes them more easily detectable. Among the stones (both falls and finds), the chondrites make up the overwhelming proportion (Table 1). Brown (1961) estimated that

Table 1. World's meteorite falls and finds from 1492-1977 (based on data from Hutchison et al., 1977, but excluding doubtful and unclassified meteorites)

Class	Falls Number	Falls Frequency (%)	Finds Number	Finds Frequency (%)
Achondrites	66	8.6	15	1.3
Carbonaceous chondrites	33	4.3	8	0.7
All other chondrites	612	80.0	516	43.8
All stones	711	92.9	539	45.7
Stony irons	11	1.4	64	5.4
Irons	43	5.6	576	48.8
All meteorites	765	100.0	1179	100.0

about 500 meteorites fall on the Earth per year, although a single fall may produce thousands of individual specimens due to break-up during entry into the Earth's atmosphere (the Pultusk, Poland stone meteorite "shower", for example, is estimated to have yielded about 100,000 individual pieces). More recent estimates indicate higher annual meteorite infall rates of about 7,500 to 200,000 for objects of an initial mass of 1 kg or greater (e.g., Vedder, 1966; McCrosky, 1968; Latham et al., 1972; Huss, 1978). However, since about 75% of the Earth's surface is covered by oceans and other vast areas are desolate and uninhabited, only a much smaller number of meteorites are actually recovered per year. From about 1800 to today, an average of approximately 4 falls and 6 finds were recovered annually the world over, although the rate of recovery has increased in recent years, due to renewed interest in, and awareness of, meteorites.

Meteorites have been known to man for thousands of years, but the oldest observed meteorite fall from which material is still preserved is the Ensisheim, Alsace, France stone which fell on November 16, 1492. However, the true nature and importance of meteorites was not recognized until 300 years later when Chladni (1794) published his revolutionary book, presenting evidence that meteorites are objects that fall to Earth from space. E.F.F. Chladni is therefore rightfully considered as the founder of the science dealing with meteorites (i.e., meteoritics). However, Chladni's (1794) ideas were not immediately accepted by his contemporaries and, in fact, he was severely criticized by some. The well documented and thoroughly investigated fall of L'Aigle (France) in 1803 dispelled the criticism, however, and since then meteorites have become the treasured objects of extensive scientific study.

Meteorites are scientifically unique objects and have therefore been intensely studied by scientists from many different disciplines, including mineralogy, petrology, geochemistry, metallurgy, inorganic and organic chemistry, nuclear chemistry and nuclear physics, astronomy, and even biology and paleontology. During the classical period of meteoritics (1794 to ~ 1900), main emphasis was on the description of meteorites, their mineralogical, petrographic, and bulk chemical characterization, and the interpretation of their textures. Between the two World Wars, interest in meteorites diminished somewhat and was largely restricted to routine recording and description of new falls and finds, and only a few die-hards the world over continued their scientific study (e.g., Merrill, Farrington, Nininger in the U.S.A., Heide in Germany). After ~ 1950, a new and exciting period of meteoritics began, largely stimulated by theoretical and experimental progress in chemistry and physics and the recognition that meteorites contain the record of the early history of our solar system, as well as due to the availability of new, sophisticated analytical tools (e.g., mass spectrometer for age dating and stable isotope geochemistry; neutron activation analysis for trace element studies; electron microprobe analysis for measurement of the composition of micron-sized mineral grains). Thus, new, exciting areas of research were opened up and new data were obtained which became the basis for our understanding of the origin and evolution of meteorite parent bodies and the early history of solid materials in our solar system. Since the first manned lunar landing in 1969, yet another stimulus was provided to meteoritics, largely because of the finding of vast rock provinces on the Moon showing the effects associated with impacts on planetary bodies devoid of atmospheres. Similar features such as shock brecciation, crushing, partial and complete melting, mixing and vapor fractionation are observed in meteorites but had been difficult to interpret. Now that samples from the Moon have been recognized to show similar features, comparative studies involving lunar and meteoritic samples have become very exciting.

3. SCIENTIFIC SIGNIFICANCE OF METEORITES

Meteorites are scientifically unique objects and their study has provided much of the data and ideas upon which hypotheses on the early history of the solar system are based. Here, we summarize the scientific significance of meteorites and what can be learned from their study. This list is by no means complete, and we emphasize those areas which we judge to be of particular interest.

Extraterrestrial materials – Meteorites are the only extraterrestrial material, besides lunar samples, that are available in relative abundance for study in the laboratory. Thus, they provide us with extraterrestrial material from regions in our solar system that otherwise cannot now be sampled, and this alone is sufficient reason to study their properties in detail.

The most ancient rocks – One of the most important tasks in cosmochemistry is to date the evolution of solid materials in the solar system. Although meteorites vary in their ages, depending upon their thermal histories after formation and upon the method of absolute age dating applied, many meteorite crystallization ages (i.e., ages obtained by methods employing solid radioactive parent and solid radiogenic daughter nuclides) cluster around 4.55 ± 0.05 billion years (b.y.). This is not only the case for the chondrites (Fig. 6), but also for some achondrites, such as the Angra dos Reis, Rio de Janeiro, Brazil, fassaite achondrite (Fig. 7, Table 2). In fact, four independent methods yield essentially the same age of 4.55 ± 0.05 b.y. for Angra dos Reis (Keil et al., 1976; Wasserburg et al., 1977; Lugmair and Marti, 1977), thus making this rock one of the oldest in the solar system (note that the oldest terrestrial rocks are about 3.8 b.y. of age). It follows that meteorite parent bodies must have formed at least 4.55 ± 0.05 b.y. ago, giving an age that is accepted to be very close to the age of the planets in our solar system. Therefore, meteorites represent very ancient material unavailable for study from any other source at present [note that the oldest lunar samples are predominantly $\lesssim$ 4.2. b.y. of age (e.g., Taylor, 1975), although some plagioclase relict ages of 4.5 b.y. (e.g., Huneke et al., 1973; Jessberger et al., 1974) and a dunite clast of 4.60 ± 0.09 b.y. (Albee et al., 1974) have been measured].

It is significant to stress that not only are some chondrites ~ 4.55 b.y. in age (meteorites which by some are interpreted as the products of the original condensation from the solar nebula) but,

Table 2. Four methods of absolute age dating of the Angra dos Reis, Rio de Janeiro, Brazil fassaite achondrite yield essentially the same age of 4.55 ± 0.05 b.y. Ages are given in billion years

	(1) $^{207}Pb/^{206}Pb$	(1) $^{206}Pb/^{238}U$	(1) $^{208}Pb/^{232}Th$	(2) Sm/Nd
Total meteorite	4.546 ± 0.001	4.63 ± 0.07	4.53 ± 0.10	
Whitlockite A	4.543 ± 0.001	4.61 ± 0.07	4.60 ± 0.11	
Whitlockite B	4.544 ± 0.001	4.546 ± 0.038	4.540 ± 0.053	
Various separates				4.55 ± 0.04

References:
(1) Wasserburg et al. (1977); (2) Lugmair and Marti (1977).

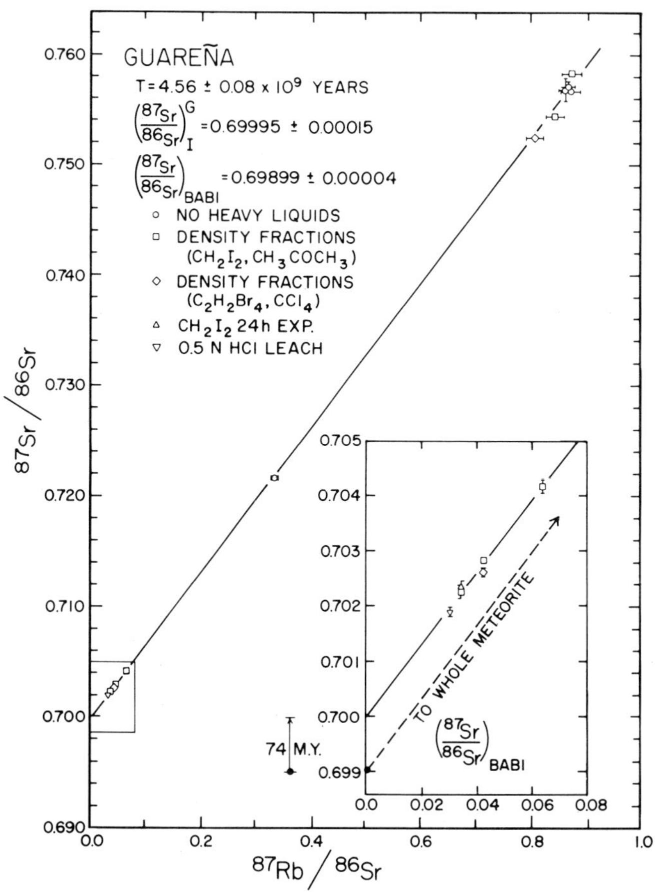

Figure 6 — Rb-Sr isochron for the Guareña, Spain, H6 chondrite giving an age of 4.56 ± 0.08 b.y. (Wasserburg et al., 1969). The $^{87}Sr/^{86}Sr$ initial value is distinctly higher than the BABI value for the eucrites and, hence, if the Rb/Sr ratio was not changed during recrystallization, that process ended ~ 74 m.y. after the fractionation of the eucrites (Reprinted with permission of Elsevier Publ. Comp.).

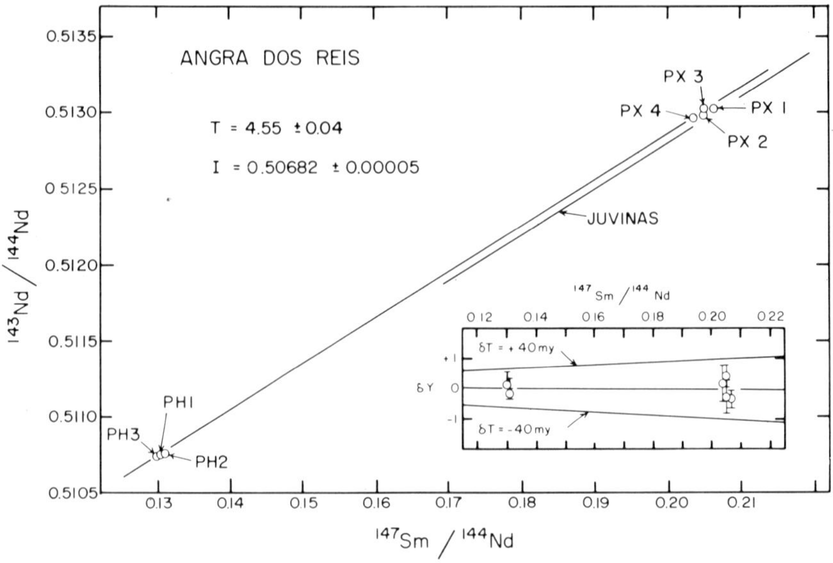

Figure 7 — Sm-Nd isochron for the Angra dos Reis, Rio de Janeiro, Brazil fassaite achondrite, corresponding to an age of 4.55 ± 0.04 b.y. Insert shows that the data are well within the ± 40 m.y. error range (Lugmair and Marti, 1977). High degree of equilibration is indicated by data for mineral separates (PH, phosphate; PX, pyroxene) clustering in two distinct, but tight sets of points (Reprinted with permission of Elsevier Publ. Comp.).

as noted above, so are some achondrites that are highly evolved igneous rocks. Angra dos Reis, for example, clearly crystallized and accumulated 4.55 ± 0.05 b.y. ago (Keil et al., 1976; Wasserburg et al., 1977; Lugmair and Marti, 1977), from a melt that must have been produced by melting or partial melting in its parent body (Prinz et al., 1977b; Ma et al., 1977). Since meteorite parent bodies are thought to be about 4.5 - 4.6 b.y. of age, at least some must have been heated to the melting points of the major silicate minerals either throughout or at least to considerable depth, during or immediately after agglomeration from the solar nebula. The heat source for this early melting is uncertain and constitutes one of the major unsolved problems in cosmochemistry. Clearly, radioactive decay of long-lived radionuclides such as those of U and K, is too slow a process to heat small meteorite parent bodies to the melting points of major rock-forming minerals in the short time-span of probably not more than 100 million years, as is indicated by the error bars on the meteorite ages (unless unreasonably older than ~ 4.6 b.y. ages are assumed for the parent bodies). Processes that have been suggested to provide the heat for this early melting are the decay of short-lived radionuclides such as ^{26}Al, but estimates of the amounts of ^{26}Al remaining at the time of formation of meteorite parent bodies indicate that they are probably too low to provide sufficient energy for melting

(Reeves and Audouze, 1968; Schramm et al., 1970). However, recent evidence and calculations suggest that abundances of ^{26}Al corresponding to an $^{26}Al/^{27}Al$ ratio of 6×10^{-5} are sufficient to heat a body of 6 km in radius to 1,800° K (zero initial temperature) (Lee et al., 1976, 1977). Another process, release of gravitational (accretional) energy as heat for mixtures of solar system abundances, appears to be insufficient for melting the small parent bodies envisaged for meteorites (Ostic, 1965). Another process was proposed by Sonett et al. (1970) who suggest that even small meteorite parent bodies can be heated to the melting points of rock-forming minerals by electric currents resulting from fields induced by high-intensity, T-Tauri-type solar wind. However, their calculations assume a number of conditions, for example, that the initial temperature must be ~ 600° K and that the solar spin rate and magnetic fields must be high. Yet another process is impact heating and melting which may have produced elevated temperatures on a local scale, but it is doubtful that it can account for the heat for all of the melting that appears to have occurred on meteorite parent bodies. Lastly, another heat source may have been the sun during the short time span (~ 10^3 years) when it was in the Hayashi state and before it became a main sequence star and when its luminosity was perhaps 500 times greater than it is today (Ezer and Cameron, 1963, 1965). Thus, it appears that none of these processes are completely satisfactory and that more research has to be awaited before the question of the heat source for the apparently nearly instantaneous melting of meteorite parent bodies immediately after (or during) their agglomeration from the solar nebula can be solved.

From nucleosynthesis to solid rock – Meteorites have also been the source of data that allow estimation of the length of the time interval between nucleosynthesis and formation of solids in the solar system, i.e., the time interval between the last formation of elements and their first incorporation into materials that agglomerated to form meteorite parent bodies. Furthermore, these data allow estimation of the length of the time intervals of formation between different meteorite types. It is concluded that the time interval between nucleosynthesis and formation of solids was probably not in excess of 100 m.y. and that formation intervals between different chondrites, achondrites, and silicate inclusions in iron meteorites were only about 14 m.y.

This dating is done on the basis of extinct radionuclides, i.e., radionuclides that have short half lives so that they are not preserved in the ancient meteorites, however, their stable decay products can be detected. This work began with the exciting discovery of Reynolds (1960a, b, 1963) that some meteorites contain excess ^{129}Xe (i.e., ^{129}Xe in excess over ambient ^{129}Xe normally found in meteorites). Reynolds (1960a, b) argues that this ^{129}Xe excess is due to the decay of now extinct ^{129}I, an isotope with a half life of 16 m.y.* It follows that excess ^{129}Xe can only have accumulated in meteorites if ^{129}I was incorporated sufficiently promptly after nucleosynthesis so that not all had decayed.** Since the half life of ^{129}I is only 16 m.y., this time interval is estimated to have been on the order of less than (and possibly considerably less than) 100 m.y. (see also, for example, Lewis and Anders, 1975; Drozd and Podosek, 1977).

As mentioned above, $^{129}I/^{127}I$ ratios inferred from Xe isotopic studies can be used to date the relative formation times between different meteorites (Fig. 8). All meteorites analyzed by Podosek (1970) appear to have formed within about 14 m.y. of each other, and even seemingly highly evolved (in the petrologic sense) rocks such as the Shallowater enstatite achondrite (a pyroxenite) and the silicate inclusions in the Campo del Cielo, Argentina, iron

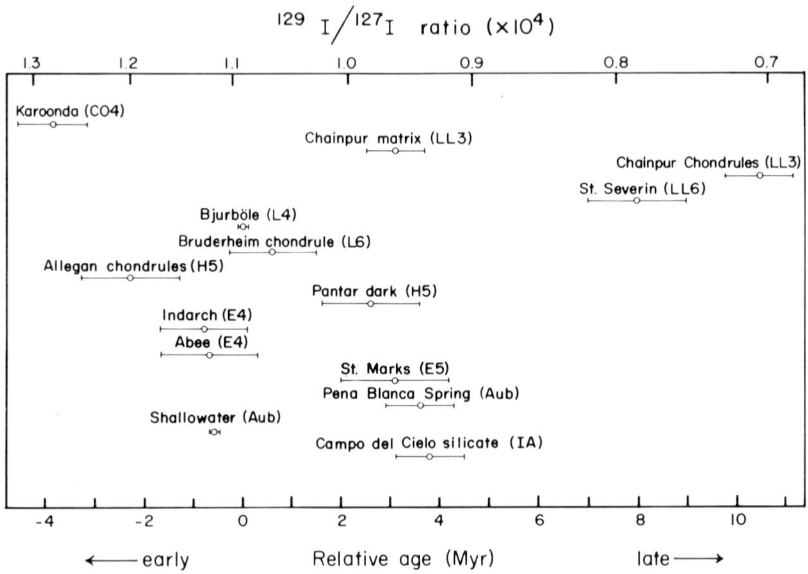

Figure 8 – Iodine-xenon formation times of meteorites, relative to the chondrite Bjurböle (Podosek, 1970). All meteorites plotted began retaining xenon at nearly the same time, within 14 m.y. of each other, i.e., they formed almost contemporaneously (Reprinted with permission of Pergamon Press).

* Since all ^{129}I has decayed long ago, the initial ^{129}I content cannot be determined and an initial $^{129}I/^{127}I$ ratio has to be assumed for all meteoritic matter. The meteorite's present ^{127}I (a stable isotope) content is taken as a measure of the former ^{129}I content and the $^{129}I/^{127}I$ ratio at time t_0 is assumed to have been 1.25×10^{-3}. Uncertainties in this ratio introduce errors into this dating technique, but this error in the ratio is judged to be less than a factor of 10, or ~ 3 half lives, or ~ ± 55 m.y. (e.g., Anders, 1964). Lewis and Anders (1975), in a more recent paper, give an $^{129}I/^{127}I$ ratio of 1.46×10^{-4} during condensation in the solar nebula.

** Note that Reynolds (1960b) and Cameron (1962) consider ^{129}I to have been the product of nucleosynthesis in the galaxy, whereas Fowler et al. (1962) suggest it to have been formed in the early solar system. The time t_0 is therefore variously interpreted as the time at which our solar system separated from the galaxy (Reynolds 1960b; Cameron, 1962) or the time at which formation of ^{129}I ceased in our solar system (Fowler et al., 1962).

meteorite, began retaining Xe at the same time as the chondrites, i.e., formed essentially contemporaneously (see also, for example, Podosek and Lewis, 1972; Drozd and Podosek, 1976).

Other evidence that the time interval between nucleosynthesis and formation of solids in our solar system was relatively short when compared to the overall old age of the solar system comes from fission products of extinct ^{244}Pu. These are manifested in fissiogenic components in the heavy xenon isotopes (e.g., Kuroda, 1960; Hohenberg et al., 1967) and in excess fission tracks (e.g., Fleischer et al., 1965). Most fission tracks (i.e., defects in crystals produced by the fission of an isotope, whereby these defects can be made visible by etching and study in the optical or electron microscope; e.g., Fleischer et al., 1975) in meteorites formed by the spontaneous or neutron-induced fission of uranium and thorium isotopes, but tracks in excess to the U and Th fission tracks have been found (e.g., Fleischer et al., 1965). These then have been attributed to the fission of ^{244}Pu which has a fission half life of

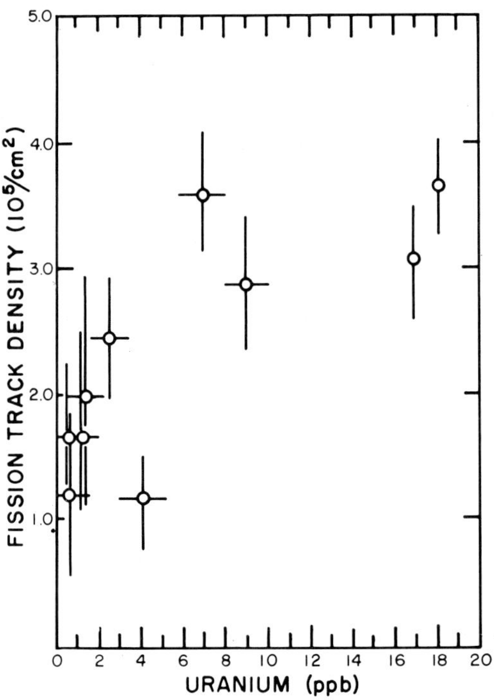

Figure 9 – Fission track density in pigeonite crystals from the Moore County, U.S.A., monomict pigeonite-plagioclase achondrite (eucrite). Total fission track density decreases by only a factor of 3 as U content decreases by a factor of 30. Thus, tracks in excess over those derived by fission of the U isotopes are interpreted as being due to the fission of extinct ^{244}Pu (after Carver, 1974).

82 m.y. (Fig. 9). Again, ^{244}Pu must have been incorporated into the solids in meteorites relatively soon after nucleosynthesis.

Kelly and Wasserburg (1978), on the basis of their discovery of excess ^{107}Ag in the Santa Clara, Mexico iron meteorite, suggested the existence of now extinct ^{107}Pd in the early solar system. This $^{107}Pd/^{107}Ag$ chronometer can measure the time interval between the last injection into our solar system of freshly synthesized radionuclides and the formation of differentiated (i.e., separated silicate-iron phase) meteorite parent bodies. This process must be commensurate with the half life of ^{107}Pd (6.5 m.y.) and, thus, these data provide further support for the proposition of a very short time interval between nucleosynthesis (injection) and meteorite parent body agglomeration and even differentiation.

Extra-solar system material — All evidence indicates that meteorites come exclusively from sources within our solar system. However, recently Clayton et al. (1973) made the exciting discovery that some of the light-colored inclusions in carbonaceous chondrites of Types 2 and 3 (HL-group) which consist predominantly of highly refractory minerals such as spinel, anorthite, gehlenite, hibonite, diopside, perovskite, grossular, etc. (see Table 6) and which are thought to represent very early,

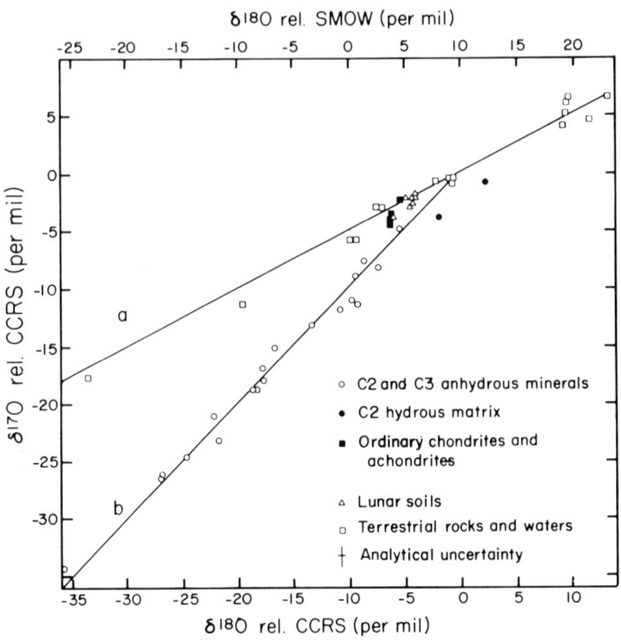

Figure 10 — Relationship between oxygen isotopes ($\delta^{17}O$ relative to carbonaceous chondrite reference standard vs. $\delta^{18}O$ relative CCRS or standard mean ocean water) in terrestrial, lunar and meteoritic samples (Clayton et al., 1973). Note that all lunar, meteoritic and terrestrial samples plot along line (a), whereas light inclusions in carbonaceous chondrites of Types 2 and 3 plot along a line (b) with dramatically different slope. The latter cannot be explained by any physical or chemical processes except nucleosynthesis (Reprinted with permission of AAAS).

high temperature condensates from the solar nebula, have nuclear oxygen isotopic anomalies that cannot be explained by any physical or chemical processes except nucleosynthesis (Fig. 10). Since it is judged unlikely that large-scale isotopic heterogeneities existed in the solar nebula from which our solar system condensed and agglomerated, Clayton et al. (1973) conclude that during agglomeration, material from outside our solar system was added to these inclusions, possibly from a supernova exploding in the vicinity of our forming solar system. Srinivan and Anders (1978) suggest ejection of dust grains from red giants as another process. Recently, nuclear isotopic anomalies were detected in carbonaceous chondrites and their light inclusions for magnesium, neon, silicon, calcium, strontium, barium, neodynium, and samarium. Anomalies for carbon, nitrogen, sulfur, krypton, mercury and uranium may also exist, but are as yet unproven. All of the anomalies are interpreted as being due to the addition of extra-solar system material (for a review of nuclear isotopic anomalies, see Clayton, 1978). Thus, these meteorites provide invaluable information on the isotopic composition of material from outside our solar system. The details of the mechanisms responsible for the addition of extra-solar system material to light inclusions are still uncertain and many exciting developments are to be expected in this field in the future.

Primitive solar system materials — A great variety of different meteorites are known, ranging from the ultramafic chondrites to petrologically evolved ultramafic, mafic and basaltic achondrites to stony-irons and irons (e.g., Mason, 1962; Keil, 1969a; Wasson, 1974). However, among the known meteorites are some that appear to be more "primitive" in elemental and isotopic abundances than any other yet analyzed material in the solar system. Thus, certain meteorites appear to represent solar system material in a very early, unmodified, "primitive" state of development and, in fact, represent the only material available to scientists that has preserved these "primitive" properties. Meteorites are therefore an invaluable scientific resource for the study of the properties of early solar system materials. This "primitive" nature is expressed, for example, by their elemental abundances. In particular, the carbonaceous chondrites of Type I have elemental abundances for the non-gaseous, relatively non-volatile elements that are very close to solar abundances (Fig. 11) (e.g., Anders, 1971b). This is a most puzzling finding because these meteorites display many petrographic indications of secondary alteration, such as veins of epsomite ($MgSO_4 \cdot 7H_2O$, possibly formerly kieserite, $MgSO_4 \cdot H_2O$) cutting through them. Furthermore, seemingly evolved ultramafic rocks such as the basaltic monomict pigeonite-plagioclase achondrites (eucrites) and the ultramafic fassaite achondrite Angra dos Reis, Rio de Janeiro, Brazil have the most primitive $^{87}Sr/^{86}Sr$ initial ratios yet measured (e.g., Papanastassiou and Wasserburg,

1969; Papanastassiou, 1970; Wasserburg et al., 1977). This is also the case for the lead isotopes, for example, of Angra dos Reis (e.g., Wasserburg et al., 1977).

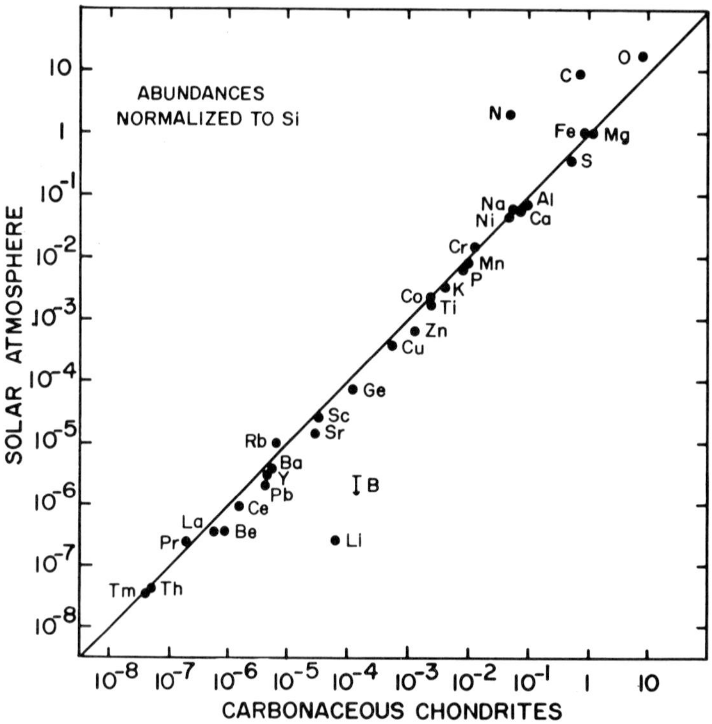

Figure 11 — Elemental abundances in carbonaceous chondrites of Type 1 plotted against solar abundances (after Ringwood, 1977, based on data by Ross and Aller, 1976), indicating that carbonaceous chondrites of Type 1, in their abundances of non-volatile, non-gaseous elements resemble abundances in the sun (Reprinted with permission of Australian National University).

Cold, dry storage — Although many chondrites are highly recrystallized (i.e., metamorphosed in the sense of Van Schmus and Wood, 1967), others have apparently been "stored" in regions in the solar system for billions of years, where reheating and other processes of secondary alteration were at a minimum, thus preserving "primitive" chemical and petrographic features. For example, the occurrence of primary igneous glass of acidic to basaltic compositions in some chondrules is a good case in point (Fig. 12). In the terrestrial environment with its relative abundance of H_2O, glasses of such compositions quickly recrystallize (note that the oldest terrestrial basaltic glasses are $\lesssim$ 1 m.y. of age). Under the dry conditions of some meteorite parent bodies, where apparently little or no reheating took place in some cases, such glasses are preserved for billions of years.

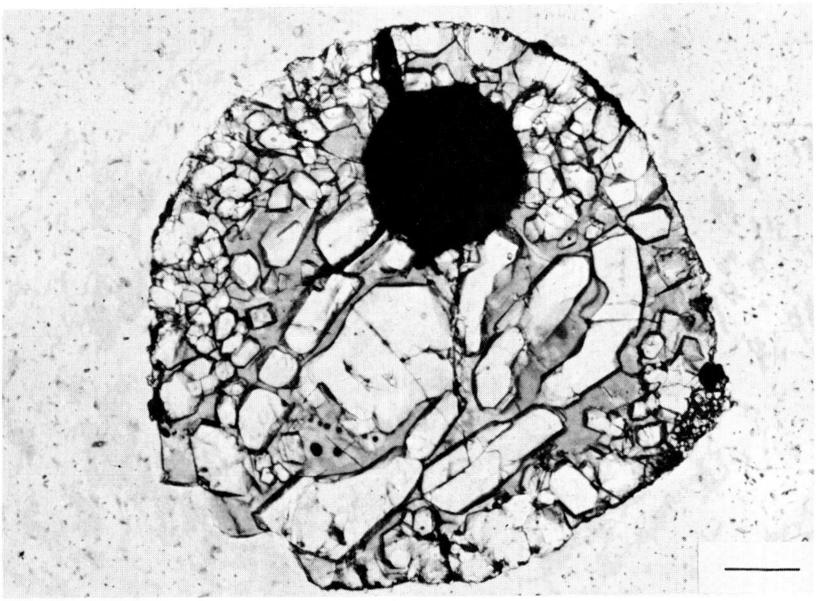

Figure 12 — Photomicrograph of a chondrule separated from the LL-3 Chainpur, India, chondrite, consisting of euhedral olivine crystals (white) embedded in a clear glass (gray). Metal droplet is black. Transmitted, plane polarized light. Scale bar equals 0.08 mm.

Cosmic ray interactions and collisions — Before their fall onto earth, meteoroids have been traveling in space as small, decimeter-to meter-sized bodies. During this journey, they have been exposed to interaction with solar wind, solar flare accelerated nuclei, cosmic and galactic cosmic rays, as well as to collisions with each other. Thus, meteorites represent "the poor man's space probe", arriving on Earth "for free" and having collected information about radiation and particle fluxes as well as collisions in space. These data, in turn, can be used, for example, to determine cosmic ray exposure ages and terrestrial ages of meteorites.

For example, cosmic rays, which consist mainly of protons (energy ~ 1 GeV), penetrate about 1 m into meteoritic matter and cause spallation-type nuclear reactions in the meteoroids which produce "cosmogenic" nuclides that can be measured after fall, if they are radioactive or otherwise distinguishable from indigeneous isotopes. If one assumes that the cosmic ray flux was constant over time and that the shape of the meteoroid was not altered, then the amount in the meteorite of a stable, cosmogenic nuclide divided by its production rate (as determined, for example, by bombarding targets with high-energy protons from an accelerator), gives the cosmic ray exposure age, i.e., the residence time in space of a meter-sized meteoroid. Measurements of such cosmic ray exposure ages for iron and H-group stone meteorites (Figs. 13, 14) indicate that iron meteorites have residence times in space that are 20-100 times greater than those of the stone meteorites. The peak around 650 ± 100 m.y. for iron meteorites indicates that apparently a major collision took place at that time that produced the IIIAB-type irons from a single object, probably from among the Mars-crossing asteroids.

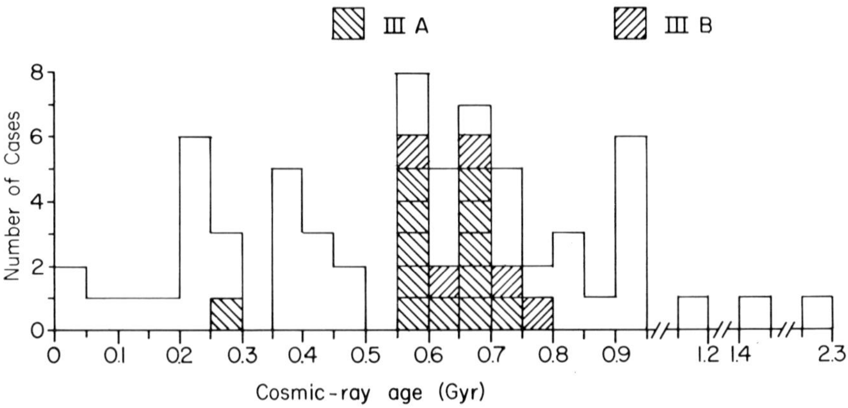

Figure 13 — Cosmic-ray exposure ages of iron meteorites, in billion years (after Wasson, 1974, based on data by Voshage, 1967). Designations III A and III B refer to specific iron meteorite groups. Irons have a residence time in space that is about 20-100 times higher than that of stones (compare to Fig. 14) (Reprinted with permission of Springer-Verlag).

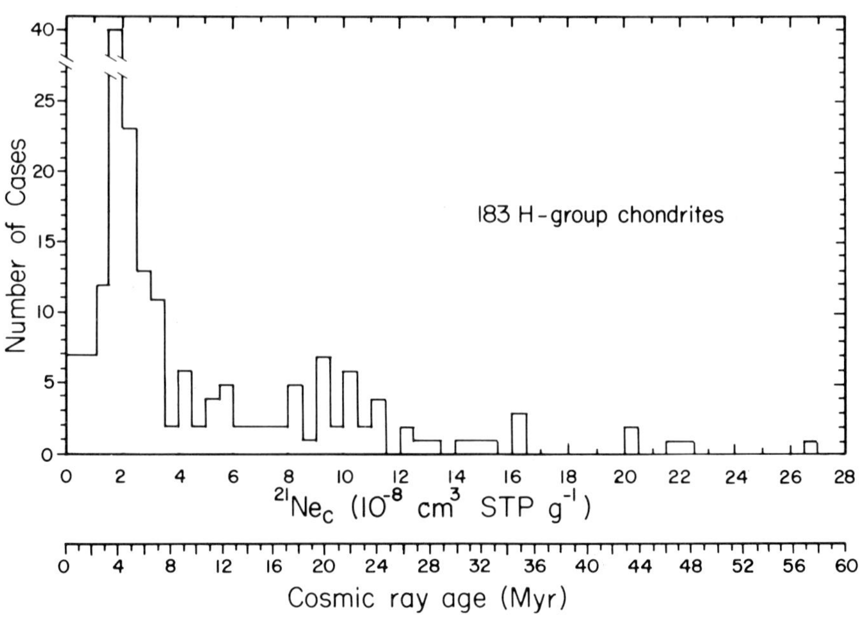

Figure 14 — Cosmic-ray exposure ages of H-group chondrites, based on their ^{21}Ne content, in million years (after Wasson, 1974, based on data by Mazor et al., 1970). Stones have a residence time in space that is about 20-100 times lower than that of iron meteorites (compare to Fig. 13) (Reprinted with permission of Springer-Verlag).

Another important parameter is the terrestrial age of meteorites, i.e., the age that a meteorite has resided on Earth. For observed falls, the date of fall provides that age. For finds for which no record of the fall is preserved the terrestrial residence time can be determined by the measurement of radioactive cosmic ray produced nuclides. After the fall of a meteorite, it is shielded from further exposure to cosmic rays by the Earth's atmosphere. Thus, radioactive cosmic ray produced nuclides will decay at a rate commensurate with their half lives, and the relative decreases in such nuclides with variable half lives provide a measure for the terrestrial age of a meteorite find. Commonly measured isotopes for this purpose are ^{36}Cl, ^{14}C, and ^{39}Ar, with half lives of 3×10^5, 5700, and 270 y., respectively. Generally, the more easily weathered stone meteorites have shorter terrestrial ages (thousands to over 20,000 y., e.g., Suess and Wänke, 1962; Boeckl, 1972) than the irons (up to 1.5 m.y., e.g., Vilcsek and Wänke, 1963). However, some of the stone meteorites recently recovered in Antarctica have very old terrestrial ages (up to 1.5 m.y., e.g., Kirsten et al., 1978), apparently because they were kept in a deep-freeze in the Antarctic ice where weathering rates are much lower than even in the arid desert environments of the world.

Regolith formation — Other important data are provided by meteorites on regolith formation on the meteorite parent bodies which presumably had no atmosphere (e.g., Housen et al., in press). A regolith is a surface layer of debris that on small and, hence, atmosphere-devoid bodies, is created by repeated impact events. The regolith on the Moon has been extensively studied (e.g., Taylor, 1975). Recent studies of meteorites indicate that some show evidence of a regolith origin, thus providing additional evidence for the impact processes that shaped the surfaces of small bodies in the solar system. The evidence for a regolith origin of certain meteorites includes, for example, the occurrence of stone meteorites exhibiting features similar to those observed in lunar breccias such as brecciation; lithic fragments formed by impact melting, partial melting and crushing; as well as the formation of agglutinates (glass-bonded aggregates) and glass spherules made by impact melting (e.g., Fredriksson and Keil, 1963; Fodor et al., 1972; Noonan, 1974; Rajan, 1974; Fodor and Keil, 1975, 1976a, b; Wilkening, 1978; Noonan et al., 1978). Further strong evidence for the regolith origin of some meteorites comes from the discovery of light primordial (solar type) noble gases in the so-called gas-rich meteorites and from the identification of gradients in solar wind induced fission tracks in individual mineral grains. Primordial rare gases occur in the dark portions of light-dark structured, brecciated chondrites such as Pantar (König et al., 1961; Fredriksson and Keil, 1963) and Tabor (König et al., 1962) and achondrites such as Kapoeta (Signer and Suess, 1963) and are attributed to incorporation as a result of irradiation by solar wind (e.g., Suess et al., 1964). Since the solar wind penetrates mineral grains only to a few microns depth, solar-type rare gases occur only near the surfaces of mineral grains (e.g., Eberhardt et al., 1965) and can only be incorporated if the mineral grains resided for some time near the surface of an atmosphere-devoid body, i.e., in a repeatedly stirred and turned-over, impact-produced regolith. Some mineral grains from the gas-rich (dark) portions of these meteorites also are rich in tracks that are thought to have been produced by solar-flare accelerated nuclei (e.g., Pellas et al., 1969). In fact, individual grains sometimes show gradients in their particle track densities, with high densities at the edge and as much as ten times fewer tracks in the center (Fig. 15). These gradients are interpreted as being due to oriented positioning of the grains near the surface of the regolith on the meteorite parent body, with the area high in tracks being closer to the surface and, thus, being exposed to a higher dose of solar--flare accelerated nuclei (e.g., Wilkening et al., 1971).

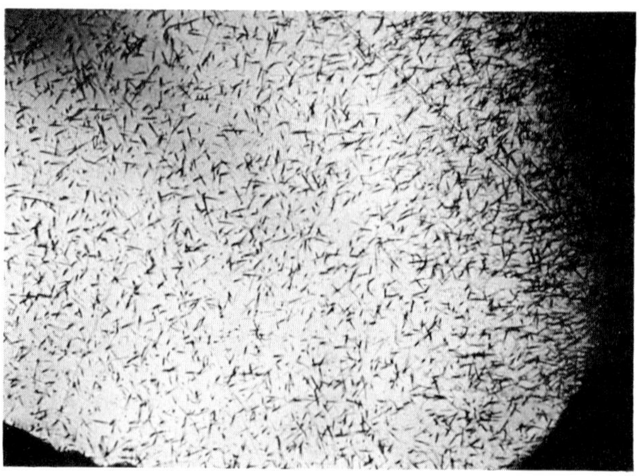

Figure 15 — Photomicrograph in transmitted light of an etched pyroxene crystal from the dark (gas-rich) portion of the Kapoeta, Sudan, polymict orthopvroxene-pigeonite-plagioclase achondrite (howardite), showing typical tracks (average ~ 8 μm long) made by solar-flare nuclei. The density of tracks is ten times higher near the surface (right) than in the interior of the grain, providing evidence for the regolith origin of the rock (photograph courtesy of L.L. Wilkening).

Sizes of parent bodies — Certain properties of meteorites also allow estimates to be made of the size(s) of the parent body(ies) in which they formed. Such estimates, for example, can be made on the basis of the cooling rates of iron meteorites with Widmanstätten structure. The Widmanstätten structure (Fig. 5) consists of an intergrowth along the faces of an octahedron of two nickel-iron alloys, kamacite (α-phase, $\lesssim$ 7% Ni) and taenite (γ-phase, $\gtrsim 7 \lesssim$ 50% Ni). This structure forms by solid state diffusion and is readily understood by examination of the Fe-Ni phase diagram (Fig. 16; Goldstein and Ogilvie, 1965a). An alloy of about 10 atomic % Ni will be entirely taenite (γ-phase) until a

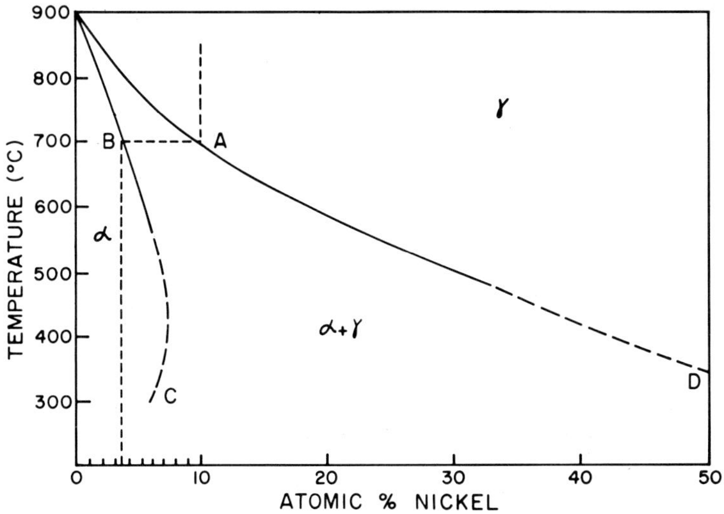

Figure 16 — Fe-Ni equilibrium phase diagram (after Goldstein and Ogilvie, 1965a) (Reprinted with permission of AIME).

temperature of about 700°C (point A) is reached. Then kamacite (α-phase) will form with a Ni-content of about 3.5 atomic % (point B). As temperature decreases, composition of γ-NiFe will change along AD and α-NiFe along BC, thus, both phases will increase in Ni. This is only possible by an increase in the amount of α-phase at the expense of the γ-phase, thus, with decreasing temperature, kamacite bands of the Widmanstätten structure increase in width whereas taenite bands decrease. At about 350°C, efficient diffusion ceases. Since cooling rates of octahedrites exceed the effective diffusion rates of Fe and Ni between kamacite and taenite, concentration gradients form in taenite and in kamacite at the taenite-kamacite interfaces. These gradients are a measure of the cooling rate and can be matched by computer-calculated gradients for known diffusion coefficients and cooling rates. Thus, the cooling rates through the temperature range of about 700–500°C of octahedrites can be determined (e.g., Wood, 1964; Goldstein and Ogilvie, 1965b; Goldstein, 1965; Goldstein and Short, 1967; Goldstein and Axon, 1973) and these are found to range between about 1–100°C/10^6 years. In turn, these cooling rates allow calculation of the minimum radii of bodies in whose very centers the octahedrites were located during cooling (or, alternatively, the burial depths in parent objects of given radii, if the octahedrites were not located in the very centers). Such calculations have to take into account the cooling rate of a point inside a parent body, depth of burial, total size of parent body, content of radionuclides (heat sources), thermal diffusivity of its material, and the initial

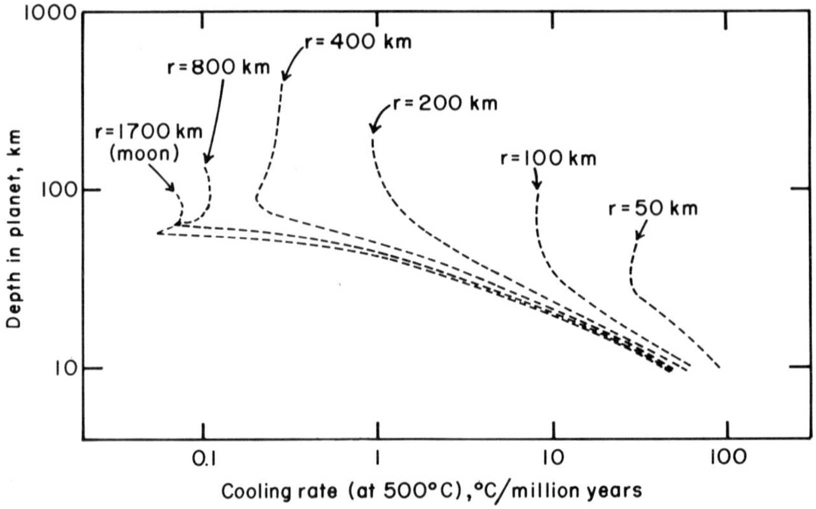

Figure 17 — Cooling rate as a function of burial depth in a meteorite parent body with chondritic radionuclide contents, an initial temperature of 1000°C, and a thermal diffusivity of 0.007 cm^2/sec at 25°C. r = radius of parent body (modified after Wood, 1967) (Reprinted with permission of Academic Press).

temperature of the object (Wood, 1967). In Fig. 17, results of such calculations for an initial temperature of 1000°C, a uniform thermal diffusivity of 0.007 cm^2/sec at 25°C, and chondritic radionuclide contents, are presented. Fig. 17 indicates that a cooling rate of $1°/10^6$ y. corresponds to a point in the very center of a parent object about 200 km in radius, whereas a cooling rate of $10°/10^6$ y. corresponds to a point in the very center of an object about 100 km in radius (alternatively, the $1°/10^6$ y. cooling rate corresponds to a point buried about 80 km in a parent body of 400 km in radius). Clearly, then, cooling rates of iron meteorites suggest that meteorite parent bodies were probably only a few hundred kilometers in radius.

Asteroidal origin — Finally, meteorites allow one to speculate as to their place of origin within our solar system. Evidence relevant to this problem comes principally from four areas, namely photography of meteorite (and fireball) falls and reconstruction of orbits; Monte-Carlo type orbital calculations; cosmic ray exposure ages and contents of trapped solar wind gases; and comparison of reflectance spectra of asteroids with those of known meteorite types.

Only three meteorite falls have so far been photographed by multiple cameras, thus allowing accurate orbit reconstruction. These are the chondrites Pribram, Czechoslovakia (Ceplecha, 1961), Lost City, U.S.A. (McCrosky et al., 1971), and Innisfree, Canada (Halliday et al., 1978). Their orbits, as well as those of many fireballs photographed by the same network of cameras, point towards an origin in the asteroidal belt, a region principally between the orbits of Mars and Jupiter that is populated by over 4000 recorded, irregularly shaped bodies (fragments) ranging in radii from 480 km (asteroid Ceres) to below the resolution of earth-based telescopes (about 0.5 km). Furthermore, careful evaluation of reports of visual observations by over 60 eyewitnesses of the June 25, 1890 Farmington, Kansas, U.S.A. chondrite fall indicate an asteroidal origin of the meteorite, specifically from among the Apollo (earth-crossing) asteroids (Levin et al., 1976).

Monte Carlo-type calculations that take into account perturbations of orbits of small bodies in the solar system by the large planets and the Sun have been carried out in an attempt to determine the ultimate source of meteorites within our solar system (e.g., Arnold, 1964, 1965a, b; Anders, 1972, 1978; Anders and Mellick, 1969; Wetherill, 1967, 1968a, b, 1969, 1974, 1977, 1978). These calculations are affected by uncertainties, particularly in the estimations of the gravitational perturbations and, thus, are only approximate. Nevertheless these calculations indicate that at least some meteorites may come from asteroidal sources, such as from

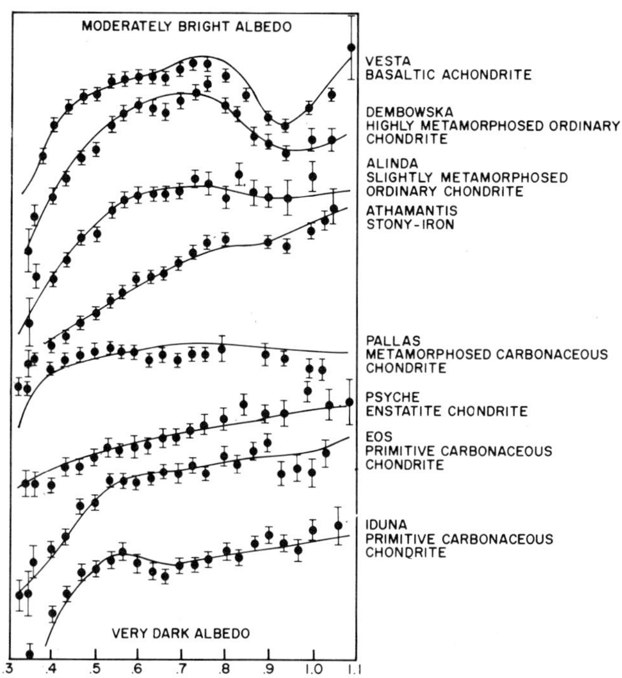

Figure 18 — Match of telescopically-determined reflectance spectra (dots) of selected asteroids (Vesta, Dembowska, Alinda, etc.) with reflectance spectra of known meteorite types obtained in the laboratory (solid lines). Spectra have been arranged in approximate order of decreasing brightness from top to bottom, and the ordinate is reflectance (vertical error bars indicate uncertainties in the telescopically-determined reflectance). The close match of asteroid with metorite spectra is powerful evidence that at least some meteorites are of asteroidal origin (after Chapman, 1975) (Reprinted with permission of Scientific American).

the Mars-crossing asteroids (e.g., Anders, 1972). The calculations also suggest that chondrites may come from Earth-crossing comets such an Encke (e.g., Wetherill, 1969). However, Anders (1975, 1978) argues that the complex textures of gas-rich meteorites and their contents of trapped solar wind gases are in conflict with a cometary but favor an asteroidal origin.

Cosmic ray exposure ages provide further evidence for the asteroidal origin of meteorites. Mean orbital lifetimes for meteoroids in Earth-crossing orbits are about 4×10^7 y. (e.g., Wetherill and Williams, 1968), whereas those for Mars-crossing orbits are about 10^9 y. (e.g., Arnold, 1964). Thus, cosmic ray exposure ages for iron meteorites (Fig. 13) are consistent with their origin from Mars- rather than from Earth-crossing asteroids.

Considerable evidence for the asteroidal origin of at least some meteorite types comes from the telescopically determined reflectance spectra of the asteroids and their comparison to spectra of known meteorite types obtained in the laboratory (e.g., McCord et al., 1970; Huguenin and McCord, 1971; Chapman et al., 1973a, b; McCord and Gaffey, 1974; Chapman, 1975, 1977; Gaffey, 1976; Gaffey and Hapke, 1978). These studies indicate that spectra obtained from different asteroids match those of, respectively, iron meteorites, stony-irons, ordinary chondrites, enstatite chondrites, achondrites (eucrites, enstatite achondrites) and carbonaceous chondrites (Fig. 18), thus providing powerful evidence for the asteroidal origin of meteorites.

4. CLASSIFICATION OF STONE METEORITES

Stone meteorites are divided into two major groups, the achondrites and the chondrites. This division is made on the basis of many parameters, but the simplest one is texture as visible in hand specimen or in the microscope: achondrites have igneous, cumulate and/or brecciated textures and resemble more closely terrestrial mafic and ultramafic rocks than any other group of meteorites (e.g., Fig. 2). Chondrites, on the other hand, contain millimeter-sized spherules called chondrules that consist mainly of silicates but also contain minor amounts of other phases (e.g., Fig. 3). Chondrules are unique to chondritic meteorites, although chondrule-like spherules of lunar impact-melt origin have also been described from lunar breccias (e.g., Keil et al., 1971, 1973; Kurat et al., 1972, 1974; King et al., 1972; Prinz et al., 1977a). Textures indicate that many chondrules formed by spontaneous crystallization of freely-floating, highly-supercooled droplets of melt (e.g., Nelson et al., 1972; Keil et al., 1973; Blander et al., 1976). Metallic nickel-iron content is another parameter of classification: achondrites usually have $< 1\%$, whereas chondrites usually have $>> 1\%$ (note that exceptions exist, for example, the highly oxidized carbonaceous chondrites of Type 1 have no metallic nickel-iron). Achondrites are further subdivided on the basis of, for example, the bulk CaO content into Ca-poor and Ca-rich achondrites. Still further subdivisions into classes are made on the basis of mineralogical, bulk compositional and textural parameters (Table 3). The Ca-poor achondrites are, thus, subdivided into the enstatite, bronzite, olivine, and olivine-pigeonite achondrites and the Ca-rich achondrites into the fassaite, augite-olivine, polymict orthopyroxene-pigeonite-plagioclase and monomict pigeonite-plagioclase achondrites. Chondrites are further subdivided into enstatite, forsterite, H-group, L-group, LL-group, HL-group, carbonaceous Type 2 and carbonaceous Type 1 chondrites. Chemical and mineralogical parameters used for the division of the "ordinary" equilibrated chondrites into H-, L-, and LL-groups are listed in Table 4. Ordinary chondrites exhibit various degrees of recrystallization, ranging from those with well-developed chondritic textures (e.g., Fig. 3) to those in which solid state recrystallization has destroyed the chondritic texture to a point where only relict chondrules are preserved (e.g., Figs. 54, 59). Concomitant to changes in the overall chondritic texture of these meteorites are changes in the degree of variability of olivine and pyroxene compositions, the structural state of low-Ca pyroxene, the degree of

Table 3. Classification of stone meteorite (modified after Keil, 1969a)*

Group	Class	Class symbol	Other names	Number of falls and finds**
Ca-poor achondrites	Enstatite achondrites	Ae	Aubrites, bustites	11
	Bronzite achondrites	Ab	Diogenites, hypersthene achondrites	10
	Olivine achondrites	Ao	Chassignites	2
	Olivine-pigeonite achondrites	Aop	Ureilites*	10
Ca-rich achondrites	Fassaite achondrite	Af	Angrite, augite achondrite	1
	Augite-olivine achondrites	Aao	Nakhlites, diopside-olivine achondrites	3
	Polymict orthopyroxene-pigeonite-plagioclase achondrites	Aor	Howardites	20
	Monomict pigeonite-plagioclase achondrites	Ap	Eucrites	30
Chondrites	Enstatite chondrites	Type I: Ce_1	Daniel's Kuil type chondrites, petrologic type 4	6
		Intermediate Type: Ce_1	Petrologic type 5	3
		Type II: Ce_2	Daniel's Kuil type chondrites, Group 1 chondrites, petrologic type 6	9
	Forsterite chondrites***	Cf		4
	High iron (H)-group chondrites	CH	Bronzite-olivine chondrites, Cronstad-type chondrites, Group 2 chondrites	517
	Low iron (L)-group chondrites	CL	Hypersthene-olivine chondrites, Baroti-type chondrites, Group 3 chondrites	515
	Low iron-low metal (LL)-group chondrites	CLL	Sometimes classified with hypersthene-olivine and Soko-Banja-group chondrites, Group 4 chondrites, amphoterites	73
	High iron-low metal (HL)-group chondrites	Vigarano-subgroup CHL-V	Pigeonite-olivine chondrites, ornansites, Group 5a chondrites, type 3 carbonaceous	18
		Ornans - subgroup CHL-O	chondrites, type 2 carbonaceous chondrites. Earlier classified with Soko-Banja-group and hypersthene-olivine chondrites	11
	Carbonaceous chondrites Type 2	Cc2	Group 5b chondrites, earlier classified with Soko-Banja group and hypersthene-olivine chondrites. Type 1 carbonaceous chondrites	18
	Carbonaceous chondrites Type 1	Cc1	Group 5c chondrites, earlier classified with Soko-Banjo group and hypersthene-olivine chondrites	5

* Among the meteorites collected during the 1977-78 joint USA — Japanese expedition to Antarctica is one achondrite weighing 482.5 grams that may constitute a new achondrite type. It consists (in vol. %) of about 55 olivine (Fa_{28}), 35 pyroxene ($En_{72}Fs_{23}Wo_5$), 8 maskelynite (An_{53}), and 2 opaques (mostly chromite, some ilmenite, traces of troilite and metallic nickel-iron) (Mason, 1978). However, further detailed research has to be awaited before the precise classification of this achondrite into a new group can be established with certainty.

** After Hutchison et al., 1977 (not included are most of the meteorites discovered in Antarctica since 1969, since the true number of separate falls represented by the over 1,300 meteorite specimens, has not yet been established reliably).

*** Forsterite chondrites Kakangari, Mount Morris (Wisconsin), Pontlyfni and Winona appear to represent a new, distinct chondrite type with olivine of Fa_{0-10}. Many other properties distinguish them from ordinary, enstatite and carbonaceous chondrites, and they may be related to chondritic xenoliths in Cumberland Falls or silicate inclusions in Group IAB iron meteorites (e.g., Graham et al., 1977a). The stone which fell near Acapulco, Mexico in August, 1976 and has olivine of Fa_{10-12}, orthopyroxene of Fs_{11-13} and chromian diopside of $En_{51-55}Fs_{4-5}Wo_{40.5-43.5}$ may be related to the same group (Christophe Michel-Levy and Lorin, 1978).

Table 4. Classification of ordinary equilibrated chondrites on the basis of compositional and mineralogical parameters

	H-group	L-group	LL-group	References
Total Fe	25.34 – 30.88 avg. 27.52	20.55 – 23.33 avg. 21.82	19.85 – 22.55 avg. 20.87	Craig (1964)
Total Ni-Fe	14.2 – 19.8 avg. 16.8	4.4 – 11.7 avg. 7.2	3.0 – 6.0 avg. 4.3	Keil (1962a, b) (revised)
Fe/SiO_2	0.77	0.55	0.49	Van Schmus and Wood (1967)
Fe^0/Fe	0.63	0.33	0.08	Van Schmus and Wood (1967)
Fe^0/Ni	10.90	6.87	3.57	Craig (1964)
Mole % Fa in Olivine	16.9 – 20.4 avg. 18.8	22.7 – 25.6 avg. 24.6	27.5 – 30.2 avg. 28.5	Keil and Fredriksson (1964; as modified by Fodor et al., 1976)
Mole % Fs in Orthopyroxene	15.7 – 18.1 avg. 17.2	18.7 – 22.6 avg. 21.3	23.2 – 25.7 avg. 24.1	
Plagioclase	$Ab_{81.9}An_{12.3}Or_{5.8}$	$Ab_{84.2}An_{10.2}Or_{5.6}$	$Ab_{85.9}An_{10.5}Or_{3.6}$	Van Schmus and Ribbe (1968)
Chromite*				Bunch et al. (1967)
TiO_2	2.33	2.81	3.23	
Cr_2O_3	56.9	56.1	54.4	
FeO	31.2	33.0	34.5	
MnO	0.94	0.74	0.63	
MgO	2.66	1.99	1.62	
Ilmenite*				Snetsinger and Keil (1969)
FeO	40.9	41.9	44.2	
MnO	3.2	1.5	1.1	
MgO	4.1	3.3	1.8	
Specific gravity	3.50 – 3.90 avg. 3.69	3.26 – 3.75 avg. 3.52	3.44 – 3.54 avg. 3.48	Pokrzywnicki (1959) and Keil (1962b)

*Listed here are only oxides that reflect chondrite classification. For complete average analyses, see listed references.

Table 5. Compositional-petrologic classification of the ordinary chondrites (modified after Van Schmus and Wood, 1967)

Parameters of classification	Petrologic types			
	3	4	5	6
Homogeneity of olivine and pyroxene	> 5% mean deviations	< 5% mean deviations to uniform	Uniform	
Structural state of low-Ca pyroxene	Predominantly monoclinic	Abundant monoclinic crystals	Orthorhombic	
Degree of development of secondary feldspar	Absent	Predominantly as microcrystalline aggregates		Clear interstitial grains
Igneous glass	Clear and isotropic primary glass. Variable abundance	Turbid if present	Absent	
Overall texture	Very sharply defined chondrules	Well-defined chondrules	Chondrules readily delineated	Poorly defined chondrules
Texture of matrix	Opaque matrix	Transparent microcrystalline matrix	Recrystallized matrix	
Bulk carbon content	0.2-1.0%	<0.2%		

development of secondary feldspar, the presence of igneous glass, the texture of the chondrite matrix, the bulk carbon contents, etc. These changes as a function of degree of recrystallization have prompted Van Schmus and Wood (1967) to divide ordinary chondrites into four petrologic types (types 3, 4, 5, 6, with types 1 and 2 not existing; Table 5), whereby type 3 is the least recrystallized (i.e., "unequilibrated") and type 6 the most recrystallized (i.e., "equilibrated") type. Recently, Dodd et al. (1975) suggested that the Shaw meteorite represents an even more highly recrystallized chondrite of type 7. However, Fodor and Keil (1975) and Taylor et al. (1979) showed that Shaw, in fact, was melted in part, in all probability by impact on the parent body. Thus, textural alterations of this rock go beyond the extreme solid state recrystallization suggested by Dodd et al. (1975).

5. MINERALOGY AND BULK COMPOSITION OF STONE METEORITES

Stone meteorites are ultramafic to mafic igneous rocks and, thus, they principally consist of members of the olivine, pyroxene, and plagioclase mineral families (except for carbonaceous chondrites of Types 1 and 2 which contain major layer-lattice silicates). Because of the generally highly-reduced nature of stone meteorites, metallic nickel-iron and troilite are often additional major phases. However, a great number of minor and accessory phases have been described from stone meteorites, and at least 90 genuine minerals have been recognized (Table 6). Of these, 16 are not known from terrestrial rocks. In addition to the genuine, well-documented mineral occurrences listed in Table 6, a number of discredited, questionable, and poorly-characterized phases have been listed in the literature.

Among the discredited phases are chalcopyrrhotite and valleriite. Chalcopyrrhotite was described by Ramdohr (1963a), but its occurrence was questioned by Cabri (1967), although the exact nature of the phase remains unknown. Valleriite was also described by Ramdohr (1963a), but Mason (1972) suggests that it may, in fact, be mackinawite.

A great many questionable phases have been described, and considerably more work is required to confirm their occurrence in stone meteorites. Sepiolite, palygorskite, metahalloysite, hornblende, and marcasite have been described by Kerridge (1962) from carbonaceous chondrites on the basis of electron diffraction, but their occurrence has not been confirmed by other investigators. Bass (1971) describes montmorillonite from the Orgueil Type 1 carbonaceous chondrite, but this occurrence requires elucidation in view of the low bulk Al_2O_3 content of this meteorite. Richardson (1978) mentions the occurrence of hexahydrite, $MgSO_4 \cdot 6H_2O$, thenardite, Na_2SO_4, vanthoffite, $Na_6Mg(SO_4)_4$, glauberite, $Na_2Ca(SO_4)_2$, mirabilite, $Na_2SO_4 \cdot 10H_2O$, loewite, $Na_4Mg_2(SO_4) \cdot 5H_2O$ and natrojarosite, $NaFe_3(SO_4)_2 \cdot (OH)_6$ in veins in carbonaceous chondrites of Type 1, but these occurrences need substantiation and verification.

Many poorly characterized phases have been mentioned in the literature that need considerably more study before they can be accepted as meteoritic phases. For example, Vdovykin (1970) describes a Ni-Cr alloy from ureilites. Fuchs and Blander (1976) noted a high-vanadium oxide (20-27% V_2O_3) in inclusions in Allende, and Fuchs et al. (1973) describe a fibrous, anisotropic Fe-S-O phase from the Murchison Type 2 carbonaceous chondrite that may be identical to the Fe-C-S phase of Ramdohr (1963a).

Table 6. Minerals occurring in stone meteorites. Minerals marked by asterisk are not known from terrestrial rocks

Mineral class	Mineral	Chemical composition	Occurrence: major constituent of	Occurrence: minor constituent of	Comments	References
Elements	Copper	Cu		Ae; Ce_2; CH; CL; CHL		Ramdohr (1973); Keil and Fredriksson (1963); Duke and Brett (1965)
	Gold	Au		Ce_2(?)	Questionable; possibly osbornite	Ramdohr (1973)
	Kamacite	(Fe, Ni) (Ni $\lesssim$ 7.5%)	Aop; Ce_1; Ce_2; Cf; CH; CL; CLL; CHL	Ae; Ab; Aa; Aor; Ap; Cc_2		Wood (1967)
	Taenite	(Fe, Ni) (Ni ~ 8-55%)	CH; CL; CLL; CHL	Ae; Ao; Aor; Ce_1; Ce_2; Cf		
	Awaruite	Ni_3Fe		CHL		Clarke et al. (1971)
	Pt-Ir-Os-Ru-Mo-W-Ni-Fe-alloys			CHL		Wark and Lovering (1976)
	Graphite	α-C		Ae; Aop; Ce_1; Ce_2; CHL; Cc_1	Cliftonites are graphite cubes, but probably not pseudomorphs after diamond	Brett and Higgins (1967)
	Diamond	β-C		Aop	Origin by shock transformation of graphite	Urey et al. (1957); Ringwood (1960); Lipschutz (1962)
	Lonsdaleite*	C (Hexagonal diamond)		Aop	Of shock origin	Frondel and Marvin (1967); Berkley et al. (1976)
	Chaoite	C, hexagonal		Aop	Of shock origin	Vdovykin (1969)
	Sulfur	α-S		Cc_1		Du Fresne and Anders (1962)
Carbides, nitrides, phosphides, silicides	Cohenite	Fe_3C		Ce_1		Brett (1966)

Table 6. Minerals occurring in stone meteorites. Minerals marked by asterisk are not known from terrestrial rocks (cont.)

	Osbornite*	TiN		Ae; Ce_2		Bannister (1941)
	Sinoite*	Si_2N_2O		Ce_2		Andersen et al. (1964); Keil and Andersen (1965a, b)
	Schreibersite	$(Fe, Ni, Co)_3P$		Ae; Ce_1; Ce_2; Cf	Rhabdite is needle- and spindle-shaped schreibersite	Keil (1968)
	Perryite*	$(Ni, Fe)_5(Si, P)_2$		Ae; Ce_1		Reed (1968); Wai (1970)
Sulfides	Heazlewoodite	Ni_3S_2		CL	Possibly terrestrial weathering product	Larimer and Rambaldi (1976)
	Pentlandite	$(Fe, Ni)_9S_8$		Ab; Ao; CH; CL; CLL; CHL; Cc_2; Cc_1		Fredriksson and Keil (1964); Ramdohr (1973)
	Sphalerite	α-ZnS		Ce_1; Ce_2		Keil (1968)
	Chalcopyrite	$CuFeS_2$		Aao; CHL		Ramdohr (1973)
	Pyrrhotite	$Fe_{1-x}S$		Cc_1		Kerridge (1970)
	Troilite	FeS	Af; Ce_1; Ce_2, Cf; CH; CL; CLL; CHL	Ae; Ab; Ao; Áao; Aop; Aor; Ap; Cc_2	Titanium bearing in Ae and Ce	Keil and Fredriksson (1963); Keil (1968, 1969b)
	Oldhamite*	CaS		Ae; Ce_1; Ce_2		Keil (1968)
	Ferroan alabandite	$(Fe_{0.23}Mg_{0.12}Mn_{0.63}Ca_{0.007})S$		Ce_2		Keil and Fredriksson (1963); Keil (1968)
	Ferro-magnesian alabandite	$(Fe_{0.2}Mg_{0.3}Mn_{0.5})S$		Ae		
	Niningerite*	$(Fe_{0.19}Mg_{0.66}Mn_{0.14}Ca_{0.007})S$ to $(Fe_{0.52}Mg_{0.33}Mn_{0.06}Ca_{0.03})S$		Ce_1		Keil (1968); Keil and Snetsinger (1967)
	Mackinawite	FeS		CH; CL	Tetragonal	Ramdohr (1973)
	Daubréelite*	$FeCr_2S_4$		Ae; Ce_1; Ce_2; Cf		Keil (1968)
	Heideite*	$FeTi_2^{3+}S_4$		Ae		Keil and Brett (1974)
	Pyrite	FeS_2		Aao; CHL		Ramdohr (1973)
	Marcasite	FeS_2		Ao		Floran et al. (1978)
	Molybdenite	MoS_2		CHL		Fuchs and Blander (1976)

Table 6. Minerals occurring in stone meteorites. Minerals marked by asterisk are not known from terrestrial rocks (cont.)

	Djerfisherite*	$K_3(Na, Cu)(Fe, Ni)_{12}S_{14}$		Ce_1; Ae		Fuchs (1966)
Halides	Halite	NaCl		Aop		Berkley et al. (1978a)
	Sylvite	KCl		Aop		Berkley et al. (1978a)
	Lawrencite*	$FeCl_2$		Ce_1		Keil (1968)
Oxides and hydroxides	Spinel	$\sim MgAl_2O_4$		CH; CHL		Sztrokay et al. (1961)
	Hercynite	$(Fe, Mg)Al_2O_4$		Af; CHL		Marvin et al. (1970);
	Magnetite	Fe_3O_4	Cc_2	Af; Aao; Aor; Ap; CH; CLL; CHL; Cc_1	Sometimes has Ni; up to 21.9% TiO_2 in Aa	Boström and Fredriksson (1966); Prinz et al. (1977b)
	Chromite	$FeCr_2O_4$	Ao	Ae(?); Ab; Aop; Aor; Ap; CH; CL; CLL; CHL	Sometimes has considerable Mg, Mn, Al	Snetsinger et al. (1967); Bunch et al. (1967)
	Ilmenite	$FeTiO_3$		Ao; Aao; Aor; Ap; CH; CL	Geikielo-ilmenite in CHL	Buseck and Keil (1966); Snetsinger and Keil (1969); Fuchs and Blander (1976)
	Hibonite	$CaAl_{12}O_{19}$		CHL; Cc_2		Keil and Fuchs (1971)
	Perovskite	$CaTiO_3$		CHL; Cc_2		Keil et al. (1969)
	Armalcolite	$FeMgTi_2O_5$		CHL		Haggerty (1977)
	Quartz	SiO_2		Aor; Ap; Ce_1; Ce_2		Mason (1966)
	Tridymite	SiO_2		Aor; Ap; Ce_1; Ce_2		
	Cristobalite	SiO_2		Ap; Ce_1; CH		
	Rutile	TiO_2		Ao; CH; CL		Buseck and Keil (1966)
	Limonite	$FeOOH.nH_2O$		Cc_1	Claimed to be of extraterrestrial origin	Nagy et al. (1963)
	Baddeleyite	ZrO_2		Ao; Af		Floran et al. (1978); Prinz et al. (1977b)
Carbonates and sulfates	Breunnerite	$(Mg, Fe)CO_3$		Cc_1; Cc_2		Kvasha (1948); Nagy and Andersen (1964)
	Calcite	$CaCO_3$		Cc_2		Kvasha (1948)
	Dolomite	$CaMg(CO_3)_2$		Cc_1		Du Fresne and Anders (1962)

Table 6. Minerals occurring in stone meteorites. Minerals marked by asterisk are not known from terrestrial rocks (cont.)

	Espsomite	$MgSO_4 . 7H_2O$	Cc_1	Cc_2	Origin of water uncertain	Du Fresne and Anders (1962)
	Astrakhanite	$Na_2Mg(SO_4)_2 . 4H_2O$		Cc_1	Origin of water uncertain	Boström and Fredriksson (1966)
	Gypsum	$CaSO_4 . 2H_2O$		Cc_1		Nagy and Andersen (1964)
Phosphates	Merrillite	β-$Ca_3(PO_4)_2$		Af; Ap; CH; CL; Cc_1 (?)		Dowty (1977)
	Chlorapatite	$Ca_5(PO_4)_3Cl$		Ao; Aao; CH; CL; CLL		Paar et al. (1976)
	Fe-brianite*	$Na_2FeMg(PO_4)_2$		CHL		Wlotzka (1976)
	Fe-panethite*	$NaFePO_4$		CHL		Wlotzka (1976)
Silicates	Forsterite-fayalite	$Mg_2SiO_4 - Fe_2SiO_4$	Ae; Ao; Aop; Aa; Aao; Cf; CH; CL; CLL; CHL; Cc_2	Ab; Aor; Ce_1; Cc_1	often ~ 0.5% Mn	Keil and Fredriksson (1964); Keil et al. (1964)
	Ringwoodite*	$(Mg, Fe)_2 SiO_4$ spinel structure		CL		Binns et al. (1969)
	Majorite*	$Mg_3(MgSi) Si_3O_{12}$ garnet structure		CL		Smith and Mason (1970)
	Monticellite	$CaMg(SiO_4)$		CH	Ferroan variety	Dodd (1971)
	Kirschsteinite	$CaFe(SiO_4)$		Af	Magnesian variety	Prinz et al. (1977b)
	Grossular	$Ca_3Al_2(SiO_4)_3$		CHL		Clarke et al. (1971)
	Andradite	$Ca_3Fe_2(SiO_4)_3$		CHL		Fuchs (1971)
	Zircon	$ZrSiO_4$		Ap		Prinz et al. (1977c)
	Gehlenite	$Ca_2Al_2SiO_7$		CHL		Keil et al. (1969)
	Melilite	$Ca_2(Mg, Al) (Si, Al)_2O_7$		CHL		Clarke et al. (1971)
	Cordierite	$Mg_2Al_4Si_5O_{18}$		CHL		Fuchs (1969a)
	Merrihueite*	$(K, Na)_2 (Fe, Mg)_5 Si_{12} O_{30}$		CL	K dominates over Na	Dodd et al. (1965)
	Roedderite*	$(Na, K)_2 (Fe, Mg)_5Si_{12}O_{30}$		Ce_1	Na dominates over K	Fuchs et al. (1966)
	Clinoenstatite-clinoferrosilite (including pigeonite)	$Mg_2Si_2O_6 - Fe_2Si_2O_6 - Ca_2Si_2O_6$	Ab; Aop; Aor; Ap; Ce_1; CH; CL; CHL; Cc_2	Aao; Ce_2; CLL		Mason (1966)
	Diopside	$CaMgSi_2O_6$		Ae; Cf; CH; CL; CHL	Including Cr-bearing endiopside	

Table 6. Minerals occurring in stone meteorites. Minerals marked by asterisk are not known from terrestrial rocks (cont.)

	Augite	$(Ca, Mg, Fe^{2+}, Al)_2(Si, Al)_2O_6$	Aao	Aop; Aor; Ap		Fredriksson and Keil (1963)
	Fassaite	$Ca(Mg, Fe, Ti, Al)(AlSi_2)O_6$	Af	CH; CHL	Fe-free in CHL; 6.7% FeO in Aa	Clarke et al. (1971); Dodd (1971); Keil et al. (1976); Prinz et al. (1977b)
	Enstatite-orthoferro-silite	$Mg_2Si_2O_6 - Fe_2Si_2O_6$	Ae; Ab; Aor; Ap; Ce_2; Cf; CH; CL; CLL; CHL; Cc_2	Ao; Ce_1	Ca ≤ 0.5%; often ~ 0.5% Mn	Keil and Fredriksson (1964)
	Kaersutite	$(K, Na)Ca_2(Mg, Fe^{2+})_4Ti(Si_6Al_2O_{22})(OH, F)_2$		Ao	Water bearing	Floran et al. (1978)
	Richterite	$Na_2Ca(Mg, Fe)_5Si_8O_{22}(F)_2$		Ae; Ce_1	Anhydrous	Olsen et al. (1973); Graham et al. (1977b)
	Rhönite	$CaMg_2TiAl_2SiO_{10}$		CHL		Fuchs (1971)
	Wollastonite	$CaSiO_3$		CHL		Fuchs (1971)
	Layer-lattice silicate (chlorite-like)	$\sim(Mg, Fe^{2+}, Ni, Al)_{5.93}OH_{7.86}O_{0.14}Si_4O_{10}$	Cc_1; Cc_2			Boström and Fredriksson (1966)
	Cronstedtite	$Fe_4^{2+}Fe_2^{3+}[(OH)_8/Fe_2^{3+}Si_2O_{10}]$	Cc_2			Müller et al. (1977)
	Nepheline	$NaAlSiO_4$		CH; CL; CLL; CHL		Kurat (1967); Fuchs (1969b)
	Potash feldspar	$(K, Na)AlSi_3O_8$		Ao; Aop; Aao		Floran et al. (1978)
	Plagioclase	$NaAlSi_3O_8 - CaAl_2Si_2O_8$	Ae; Ap; Ce_2; CH; CL; CLL; CHL	Ab; Ao; Aop; Af; Aao; Aor; Ce_1; Cf	Little K. Maskelynite is shock transformed plagioclase glass; anorthite in CHL:	Mason (1965)
	Celsian	$BaAl_2Si_2O_8$		Af	$Cs_{90.2}An_{7.7}Ab_{1.7}Or_{0.4}$	Prinz et al. (1977b)
	Sodalite	$Na_8Al_6Si_6O_{24}Cl_2$		CHL		Clarke et al. (1971)
Organic substances	Whewellite	$CaC_2O_4.H_2O$		Cc_2		Fuchs et al. (1970)

In addition, minerals A, B, and D through L of Ramdohr (1963a) and Mg_2TiO_4 of Ramdohr (1963b) require detailed study. Other poorly characterized phases include brucite or periclase from Orgueil (Boström and Fredriksson, 1966); a V_2O_3-rich phase and $Ca_{0.55}Mg_{0.32}Mn_{0.10}Fe_{0.02}CO_3$ from a carbonaceous lithic fragment in the Supuhee H6 chondrite (Leitch and Grossman, 1976); $Fe_{0.54}Zn_{0.25}Mn_{0.16}Mg_{0.004}S$ in the Yilmia enstatite chondrite (Buseck and Holdsworth, 1972); a new phase broadly related to zirconolite ($CaZrTi_2O_7$) in the Allende H L-group chondrite (Haggerty, 1977); a silico-phosphate from the Rose City H-group chondrite (Fruland and Walton, 1976); and complex, highly-variable alloys of Mo-W-Pt-Ru-Os-Ir in light inclusions from the Allende chondrite (e.g., Wark and Lovering, 1976).

Finally, not included in Table 6 are iddingsite (a mixture) described from the augite-olivine achondrites (Bunch and Reid, 1975; Berkley et al., 1979), igneous acidic glass (e.g., Fredriksson and Reid, 1965), organic compounds (e.g., Nagy, 1975), and secondary minerals of terrestrial weathering origin (e.g., Mason, 1967b).

Below are listed the major, minor and accessory phases in stone meteorites (modified after Keil, 1969a).

Achondrites — *Enstatite achondrites* — Major minerals: enstatite (< 0.1% FeO), forsterite (< 0.1% FeO), plagioclase. Minor and accessory minerals: kamacite, taenite, ferromagnesian and ferroan alabandite, schreibersite, rhabdite, titanium-bearing troilite, heideite, daubréelite, graphite, native copper, osbornite, djerfisherite, perryite, diopside, richterite.

Bronzite achondrites — Major minerals: bronzite, clinobronzite. Minor and accessory minerals: plagioclase, olivine, chromite, troilite, pentlandite, kamacite.

Olivine achondrites — Major minerals: olivine (~ 95%), chromite. Minor and accessory minerals: pyroxene, plagioclase, taenite, pentlandite, troilite, marcasite, ilmenite, rutile, baddeleyite, chlorapatite, kaersutite, potash feldspar.

Olivine-pigeonite achondrites — Major minerals: olivine, pigeonite, kamacite. Minor and accessory minerals: troilite, graphite, diamond, lonsdaleite, chaoite, halite, sylvite, chromite, augite, andesine, K-feldspar.

Fassaite achondrite — Major minerals: > 90% fassaite, olivine. Minor and accessory minerals: hercynite, baddeleyite, whitlockite, magnesian kirschsteinite, plagioclase, celsian, metallic nickel-iron, titanomagnetite, troilite.

Augite-olivine achondrites — Major minerals: augite, olivine. Minor and accessory minerals: titanomagnetite, plagioclase, alkali feldspar, ferropigeonite to ferroaugite, ilmenite, pyrite, troilite, chalcopyrite, chlorapatite, free SiO_2, glass.

Polymict orthopyroxene-pigeonite-plagioclase achondrites — Major minerals: bronzite to ferrohypersthene, pigeonite. Minor and accessory minerals: olivine, feldspar, augite, kamacite, taenite, troilite, chromite, magnetite, tridymite, quartz, ilmenite, glass.

Monomict pigeonite-plagioclase achondrites — Major minerals: pigeonite, orthopyroxene, plagioclase. Minor and accessory minerals: magnetite, ilmenite, chromite, quartz, cristobalite, tridymite, maskelynite, whitlockite, kamacite, troilite, zircon, augite.

Chondrites — *Enstatite chondrites of Type I* — Major minerals: clinoenstatite, kamacite, troilite. Minor and accessory minerals: taenite, graphite, oldhamite, daubréelite (including zincian variety), niningerite, sphalerite, djerfisherite, enstatite, albite, olivine, roedderite, quartz, tridymite, cristobalite, schreibersite, cohenite, lawrencite, perryite.

Enstatite chondrites of Type II — Major minerals: enstatite, kamacite, troilite, plagioclase. Minor and accessory minerals: taenite, native copper, graphite, cohenite, oldhamite, daubréelite, schreibersite, ferroan alabandite, sphalerite, quartz, tridymite, clinoenstatite, sinoite, gold (?).

Forsterite chondrites — Major minerals: forsterite, enstatite, kamacite, troilite. Minor and accessory minerals: daubréelite, schreibersite, plagioclase, diopside, taenite.

H-group chondrites — Major minerals: olivine, bronzite, clinopyroxene, plagioclase (or maskelynite), kamacite, taenite, troilite. Minor and accessory minerals: chromite, spinel, ilmenite, rutile, cristobalite, magnetite, chlorapatite, whitlockite, mackinawite, native copper, pentlandite, monticellite, diopside, fassaite, nepheline.

L-group chondrites — Major minerals: olivine, bronzite, clinopyroxene, plagioclase (or maskelynite), kamacite, taenite, troilite. Minor and accessory minerals: diopside, chromite, ilmenite, rutile, titanomagnetite, chlorapatite, whitlockite, native copper, mackinawite, merrihueite, heazlewoodite, pentlandite, ringwoodite, majorite, nepheline.

LL-group chondrites — Major minerals: olivine, bronzite, plagioclase, kamacite, taenite, troilite. Minor and accessory minerals: clinopyroxene, chromite, chlorapatite, magnetite, pentlandite, nepheline.

HL-group chondrites — Major minerals: olivine, clinopyroxene, pigeonite, orthopyroxene, plagioclase, troilite and/or pentlandite, kamacite, taenite, glass. Minor and accessory phases: magnetite, chalcopyrite, pyrite, spinel, graphite, native copper, awaruite, Pt-Ir-Os-Ru-Mo-W-Ni-Fe alloys, molybdenite, hercynite, chromite, hibonite, perovskite, armalcolite, Fe-brianite, Fe-panethite, grossular, andradite, gehlenite, melilite, cordierite, diopside, fassaite, rhönite, wollastonite, nepheline, sodalite.

Carbonaceous chondrites of Type 2 — Major minerals: layer-lattice silicate, olivine, pyroxene, magnetite, cronstedtite. Minor and accessory minerals: troilite, pentlandite, epsomite, calcite, kamacite, breunnerite, glass, carbon, organic compounds, hibonite, perovskite, whewellite.

Carbonaceous chondrites of Type 1 — Major minerals: layer-lattice silicate, epsomite. Minor and accessory minerals: native sulfur, gypsum, astrakhanite, breunnerite, dolomite, limonite, whitlockite, olivine, graphite, organic compounds, pentlandite, pyrrhotite, magnetite.

Textures of stone meteorites are varied: achondrites exhibit a wide range of cumulate (Figs. 2, 20, 23) to basaltic textures. Frequently, achondrites and particularly the basaltic achondrites (i.e., the polymict orthopyroxene-pigeonite-plagioclase and the monomict pigeonite-plagioclase achondrites) are brecciated, in some cases clearly due to prolonged residence in regoliths on their parent bodies. Chondrites display textures typical of tuffaceous agglomerates (i.e., with chondrules, chondrule and mineral fragments embedded into a varied, usually finer-grained matrix; Fig. 3), with differing degrees of recrystallization having altered the original texture (e.g., Figs. 3, 45, 50, 54). As is the case with some achondrites, some chondrites show brecciation due to residence in parent body regoliths.

The bulk major and minor element compositions of stone meteorites, analogous to the mineralogical composition, reflect the ultramafic to mafic nature of these rocks (Table 7): analyses are generally low in SiO_2, Al_2O_3, CaO and the alkalis, and high in MgO. Total iron (i.e., combined iron occurring as Fe^0 and Fe^{2+} in metal, silicates, troilite, chromite, ilmenite, etc.) is usually high, and metallic nickel-iron is also frequently abundant.

Table 7. Average major and minor element compositions of stone meteorites, in weight percent (modified and updated after Keil, 1969a)

Class	Silicate portion															Metal portion									
	SiO_2	TiO_2	Al_2O_3	Cr_2O_3	Fe_2O_3	FeO	MnO	MgO	CaO	NiO	Na_2O	K_2O	P_2O_5	H_2O	total silicate portion	Fe	Ni	Co	P	Total metal portion	FeS	C	Others	Number of meteorites averaged	References
Enstatite achondrites	54.01	0.06	0.67	0.06		0.97	0.14	35.92	0.91	0.26*	1.32	0.10	0.22	1.14	95.78	2.29	0.17			2.46	1.25		0.51**	4	Urey and Craig (1953) and Wiik (1956), weighted equally
Bronzite achondrites	52.11	0.19	1.18	0.80		16.05	0.32	25.85	1.41		0.004	0.001	0.01	0.14	98.06	0.79	0.03			0.82	1.12			4	Urey and Craig (1953)
Olivine achondrites	37.36	0.10	1.02	0.79		26.51	0.46	31.47	1.11		0.23	0.06	0.11	0.24	99.46		***	***			***	***		2	Calculated from Table 4 of Floran et al. (1978)
Olivine-pigeonite achondrites	40.54	0.14	0.63	0.79		13.51	0.37	36.69	0.95		0.10	0.04	0.09		93.85	3.39	0.11	0.05		3.55	1.42	2.23		3	Wiik (1969) Dyakonova (1964)
Fassaite achondrite	43.94	2.39	8.73		0.31	8.28		10.05	24.51		0.036+	0.0049+	0.13		98.38						1.26			1	Ludwig and Tschermak (1887, 1909)
Augite-olivine achondrites	48.88	0.32	1.60	0.32	1.22	19.13	0.61	11.70	15.45		0.62	0.27	0.12		100.24						0.06(S)			2	Prior (1912), McCarthy et al. (1974), Burragato et al. (1975)
Polymict ortho-pyroxene-pigeonite-plagioclase achondrites	49.75	0.11	8.71	0.42	2.62	13.26	0.78	16.10	6.53		0.95	0.28	0.07	0.25	99.83	0.33	0.11			0.44	0.73			6	Urey and Craig (1953)
Monomict pigeonite-plagioclase achondrites	48.95	0.70	12.58	0.34		18.76	0.56	7.16	10.33		0.53	0.043	0.096		100.05									4	McCarthy et al. (1973)
Enstatite chondrites Type I	35.81	Ti = 0.07	1.95	Cr = 0.24			Mn = 0.18	17.90	1.43		0.93	0.13				total Fe = 32.85	1.77	0.09	0.17		S = 5.77	0.43#		5	Mason (1966)
Enstatite chondrites Type II	41.15	Ti = 0.05	1.98	Cr = 0.29			Mn = 0.10	22.89	1.24		0.85	0.14				total Fe = 24.88	1.53	0.11	0.12		S = 3.31	0.34#		6	Mason (1966)
Forsterite chondrites	35.38	0.09	2.35	0.24	++	2.74	0.17	23.18	1.42	0.32	0.86	0.08	0.18	1.85	68.86	13.56	1.38	0.06	0.08θ	15.08	15.48θθ	0.50		3	Graham et al. (1977a)
High iron (H)-group chondrites	36.52	0.13	2.43	0.36		8.87	0.25	23.48	1.82		0.85	0.14	0.23	0.33	75.42	17.23	1.58	0.085		18.90	5.35	0.10#		27	Craig (1964)
Low iron (L)-group chondrites	39.88	0.15	2.31	0.44		13.12	0.27	24.98	1.90		0.88	0.14	0.26	0.34	84.67	7.70	1.12	0.059		8.88	6.17	0.09#		30	Craig (1964)
Low iron-low metal (LL)-group chondrites	39.38	0.19	2.21	0.30		17.70	0.32	25.40	1.96		0.87	0.21	0.22	0.37	89.58	3.39	0.95	0.052		4.39	5.87			7	Craig (1964)
High iron-low metal (HL)-group chondrites	33.88	0.13	2.67	0.50		24.00	0.20	23.68	2.25		0.59	0.049	0.30	0.53	88.78	3.64	1.43	0.007		5.15	5.90	0.33		4	Craig (1964)
Carbonaceous chondrites 2	27.31	0.10	2.31	0.39		20.06	0.17	19.00	2.03	1.56	0.54	0.05	0.27	13.23	87.02	0.00	0.16	0.00		0.16	8.58θθθ	2.44	1.80†	8	Wiik (1956)
Carbonaceous chondrites 1	23.08	0.08	1.77	0.28		10.32	0.19	15.56	1.51	1.17	0.76	0.07	0.27	20.54	75.60	0.11	0.02	0.00		0.13	16.88θθθ	3.62	3.77†	3	Wiik (1956)

* Very doubtful, due to the highly reduced nature of enstatite achondrites.

** CaS according to the analyses. These analyses are incorrect inasmuch as enstatite achondrites contain minor amounts of Fe and S in $FeCr_2S_4$; Fe,Mg,Mn, and S in ferroan alabandite and ferromagnesian alabandite; and most of the Ti in troilite combined with sulfur and in osbornite combined with nitrogen.

*** The analysis of Brachina (Johnson et al., 1977) includes 0.07% C, 3.59% FeS, 0.56% NiS, and 0.05% CoS.

\+ INAA analyses by Ma et al. (1977).

\+ + 0.75% Fe_2O_3 in Mt. Morris, due to terrestrial weathering.

\# After Moore and Lewis (1965, 1966).

θ Schreibersite.

θθ Contains 0.07 Mn, 0.03 Ca, and 0.06 Cr.

θθθ Contains little FeS, but most of the sulfur as elementary S, water-soluble sulfates, and pentlandite.

† Ignition loss, considered to be approximately equal to content of complex organic matter.

6. HOW TO RECOGNIZE A METEORITE

There are no single features of suspected meteorites, nor tests that can be made, that unequivocally distinguish all meteorites from terrestrial rocks and artifacts. Thus, deciding whether a suspected meteorite specimen is indeed a meteorite and not a terrestrial rock or a man-made artifact (e.g., a slag) is not always easy and requires experience and careful application of certain tests. These may involve simple visual inspection or more complicated microscopic study of thin or polished sections, chemical and isotopic analyses, and others. The most unequivocal test regarding the meteoritic nature of a specimen is the analysis for cosmic ray produced nuclides. Since terrestrial rocks are shielded from cosmic ray flux by the terrestrial atmosphere, they do not contain cosmic ray produced nuclides. Only objects that have been in space for some time as relatively small, meter-sized objects contain such nuclides, and these can be measured as long as the terrestrial residence time of the meteorite has not exceeded more than a few half lives of the particular nuclide.

Except for some achondrites and carbonaceous chondrites, meteorites formed under much more reducing conditions than did terrestrial rocks and, thus, most contain iron and nickel as native alloys (only a few terrestrial rocks contain native iron or Ni-Fe alloys). Hence, the presence of high-reflecting metallic nickel-iron specks in a silicate matrix in hand-specimen and under the microscope (e.g., Figs. 3A, 42, 49) or the chemical detection of Ni are, in most cases, excellent indicators for the stone meteorite nature of a questionable specimen. In case of iron (Fig. 1) and stony-iron (Fig. 4) meteorites, the large proportion of metallic nickel-iron, the content of nickel, and the characteristic Widmanstätten structure (Fig. 5) that formed by extremely slow cooling ($\sim$ 1-100°C/10^6 years) are excellent diagnostic features. The chondritic texture (Fig. 3) is not known from terrestrial rocks and, thus, it is also a good indicator that a specimen is a meteorite. Furthermore, since meteorites enter the terrestrial atmosphere with cosmic volocities ($\lesssim$ 40 km/sec), ablation of the specimen takes place and a characteristic, millimeter-thick, black, dull to shiny melting crust forms (e.g., Figs. 27, 36, 60) that changes to brown in color in case of stone finds due to terrestrial weathering. Although some terrestrial rocks show features which, at first sight, appear similar to the melting crust (e.g., desert varnish), the former upon closer inspection, can readily be distinguished from varnish and, thus, melting crust is a reliable indicator of the meteoritic nature of a specimen. Other ablation features such as regmaglypts (thumbmark-like surface imprints; Figs. 40, 41, 60) and aerodynamic orientation of entire specimens with, for example, conical-or dome-shaped form with ablation furrows around the periphery are also good indicators for the meteoritic origin of a specimen.

It appears, therefore, that with diligence, most specimens of suspected meteoritic origin can readily be identified and distinguished from terrestrial rocks and artifacts. Yet, it is entirely possible, if not likely, that particulary friable types have gone unnoticed and have escaped collection. Thus, the world's meteorite collections are probably incomplete and do not contain representatives of all meteoroids occurring in the interplanetary space.

7. BRAZILIAN STONE METEORITES

In this chapter, Brazilian stone meteorites will be described in detail, beginning with the achondrites and followed by the H-, L-, and LL-group chondrites (Table 8). Each meteorite is discussed in terms of fall circumstances, texture, mineralogy, chemical composition, nuclear particle tracks, age, origin, and available material.

Table 8. Historical summary and classification of Brazilian stone meteorites

Class			Fall	Find	Year	State
Achondrites	Angrite	Angra dos Reis	X		1869	RJ
	Nahklite	Governador Valadares		X	1958	MG
	Eucrites	Ibitira	X		1957	MG
		Serra de Magé	X		1923	PE
Chondrites	H-group	Avanhandava	X		1952	SP
		Conquista	X		1965	MG
		Ipiranga	X		1972	PR
		Itapicuru Mirim	X		1879	MA
		Macau	X		1836	RN
		Marília	X		1971	SP
		São José do Rio Preto	X		1962	SP
		Sete Lagoas	X		1908	MG
		Uberaba	X		1903	MG
	L-group	Mafra	X		1941	SC
		Minas Gerais		X	1888.	MG
		Paranaíba	X		1956	MT
		Patrimônio	X		1950	MG
		Putinga	X		1937	RS
		Rio Negro	X		1934	PR
		Santa Bárbara	X		1873	RS
	LL-group	Parambu	X		1964	CE

Abbreviations of the States are as follows:
RJ = Rio de Janeiro; MG = Minas Gerais; PE = Pernambuco; SP = São Paulo;
PR = Paraná; MA = Maranhão; RN = Rio Grande do Norte; SC = Santa Catarina;
MG = Mato Grosso; RS = Rio Grande do Sul; CE = Ceará.

A total of 21 stone meteorites are known from Brazil (Table 8); of these, all but two (Governador Valadares and Minas Gerais) are observed falls. As is to be expected, most of these meteorites were recovered in the highly-populated east coast states of Brazil (Fig. 19).

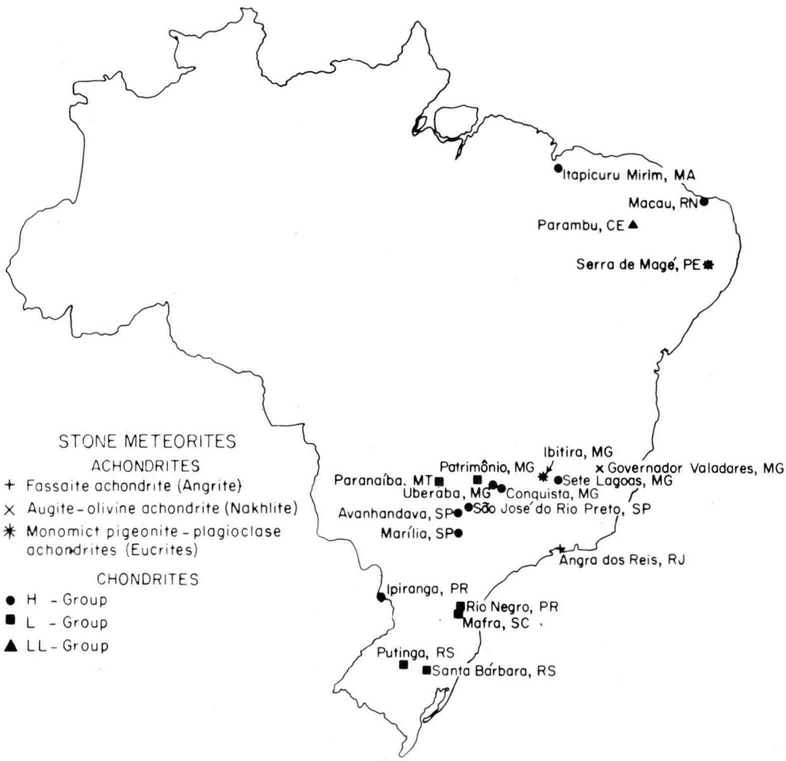

Figure 19 – Map of Brazil, showing the locations of falls and finds of stone meteorites (modified after Gomes and Keil, 1977b). Also indicated are the meteorite classifications. Most meteorites were recovered in the highly-populated east coast states of Brazil.

7.1. ACHONDRITES

7.1.1. Fassaite achondrite

Angra dos Reis (22°58'S; 44°19'W)

Fall circumstances – The Angra dos Reis, State of Rio de Janeiro, meteorite fell in the morning (5.00 hr) of an undetermined day in the latter half of January, 1869. The fall was observed by several eyewitnesses, including the local judge. The meteorite fell into the bay of Angra dos Reis at a water depth of 2 m, immediately in front of the church of Bom Fim in the town of Angra dos Reis. Some smoke was noticed in the sky and the body apparently traveled from north to south. The material, recovered by a local diver a day after the fall, consisted of two small pieces;

from an unmatched fresh surface it was assumed that a third piece was missing. Derby (1888a, b), who reported detailed field obervations of the fall, described one of these fragments as weighing 446.5 g, but no reference was made to the second one. On the other hand, Ludwig and Tschermak (1887, 1909) and Tschermak (1888) reported that the recovered material was originally one stone of about 1.5 kg. The meteorite is listed in Oliveira (1931), Vidal (1936), Hey (1966) and Hutchison et al. (1977).

Texture — The meteorite shows a granular, recrystallized texture mainly formed by small xenomorphic fassaite grains up to 0.5 mm (average 0.1 mm) in size. Less common are large (up to 3.0 mm; average 1.3 mm) poikilitic fassaite crystals that enclose the groundmass grains (Fig. 20). No evidence of strain or shock deformation has been noted. Petrofabric analysis shows a preferred orientation with weak lineation for groundmass fassaite and an apparently random orientation for the poikilitic fassaite grains, indicating a cumulate origin for the rock (Prinz et al., 1977b).

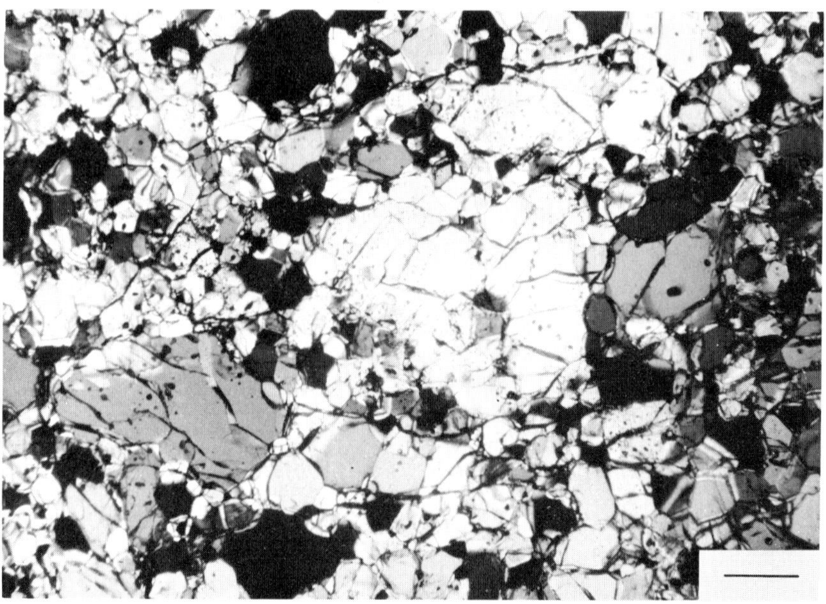

Figure 20 — Photomicrograph of the Angra dos Reis, Rio de Janeiro, Brazil fassaite achondrite, showing small, xenomorphic and large, poikilitic fassaite grains, constituting about 95% of the meteorite. Transmitted light, crossed nicols. Scale bar equals 0.30 mm.

Mineralogy — Angra dos Reis is an ultramafic igneous rock, a pyroxenite, consisting of 93% pyroxene (Ludwig and Tschermak, 1887, 1909). This mineral is slightly pleochroic, ranging in color from nearly colorless to red-brown, and has high TiO_2 and CaO contents. It is thus properly referred to as fassaite (Hutchison,

1972, 1973; Keil et al., 1976; Prinz et al., 1977b). No evidence of zoning or exsolution was found and it is homogeneous in composition, although Hazen and Finger (1977) suggest minor compositional variations. The analysis in Table 9 and Fig. 21, averaging $En_{33.5}Fs_{11.9}Wo_{54.6}$, is very similar to those published by others of fassaite of this rock (Albee and Chodos, 1970; Hutchison, 1972; Hazen and Finger, 1977; Ma et al., 1977). The endmember compositions given in Table 9 and above in terms of En, Fs and Wo do not, however, reflect the true composition of this Ti-rich pyroxene. Thus, Prinz et al. (1977b) calculated the analysis on the basis of the following endmembers: $CaMnSi_2O_6$, 0.2%; $Ca(Mg_{0.5}Ti_{0.5})(SiAl)O_6$, 9.0%; $Ca(Fe_{0.5}Ti_{0.5})(SiAl)O_6$, 3.2%; $CaCr(SiAl)O_6$, 0.6%; $CaAl(SiAl)O_6$, 15.7%; $Ca_2Si_2O_6$, 34.1%; $Mg_2Si_2O_6$, 27 4%; and $Fe_2Si_2O_6$, 9.7%. A calcium-rich olivine (Table 9; cf., Prinz et al., 1977b; see also Hutchison, 1972), occasionally containing tiny inclusions of magnesian kirschsteinite (Table 9), makes up about 5.5% of the meteorite. Accessory phases include spinel, celsian, plagioclase, whitlockite (merrillite), titano-magnetite, metallic nickel-iron, troilite, baddeleyite, SiO_2, and terrestrial hydrous iron oxide (Keil et al., 1976; Prinz et al., 1977b). The occurrence of daubréelite (Ramdohr, 1973) is doubtful in view of the relatively highly oxidized nature of this rock and could not be confirmed by Keil et al. (1976) and Prinz et al. (1977b). The

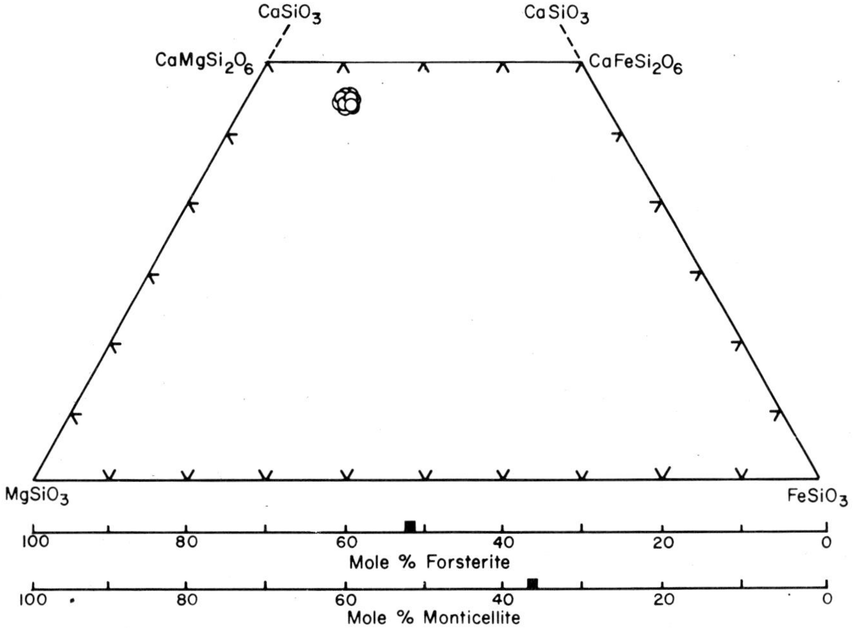

Figure 21 – Compositions of fassaite expressed in terms of molecular endmembers enstatite ($MgSiO_3$), ferrosilite ($FeSiO_3$) and wollastonite ($CaSiO_3$), for the Angra dos Reis, Rio de Janeiro, Brazil fassaite achondrite. Also shown are the compositions of coexisting olivine and magnesian kirschsteinite (Prinz et al., 1977b).

Table 9. Average compositions of silicate, oxide, and phosphate minerals in the Angra dos Reis fassaite achondrite (after Prinz et al., 1977b)

	Fassaite	Olivine	Magnesian kirschsteinite	Spinel	Plagioclase	Whitlockite	Celsian	Titano-magnetite
SiO_2	45.9	36.3	34.6	0.51	46.2	0.67	33.6	1.23
TiO_2	2.16	0.05	0.02	0.65	n.d.	n.d.	n.d.	21.9
Al_2O_3	10.0	<0.02	0.33	54.5	33.6	n.d.	28.0	3.5
Cr_2O_3	0.21	0.03	0.02	3.3	n.d.	n.d.	n.d.	1.02
V_2O_3	n.d.	n.d.	n.d.	0.07	n.d.	n.d.	n.d.	0.26
Fe_2O_3*	n.d.	n.d.	n.d.	3.4	n.d.	n.d.	n.d.	18.5
FeO	6.7	38.3	26.2	28.4	0.21	1.29	0.43	48.8
MnO	0.06	0.60	0.42	0.18	n.d.	n.d.	n.d.	1.17
MgO	10.6	24.3	8.9	8.0	n.d.	2.82	n.d.	1.18
CaO	24.1	1.29	28.9	0.76	17.2	49.4	1.20	1.21
Na_2O	<0.02	n.d.	0.05	n.d.	1.47	0.68	0.15	n.d.
K_2O	n.d.	n.d.	n.d.	n.d.	0.11	n.d.	0.05	n.d.
P_2O_5	n.d.	n.d.	n.d.	n.d.	n.d.	45.1	n.d.	n.d.
BaO	n.d.	n.d.	n.d.	n.d.	n.d.	n.d.	38.2	n.d.
TOTAL	99.73	100.87≠	99.44	99.77	98.79	100.22+	101.63	98.77≠
	En 33.5	Fo 53.1	Mo 37.7	Cm 3.8	Ab 13.3		Ab 1.7	Cm 3.8
	Fs 11.9	Fa 46.9	Ks 62.3	Hc 56.4	An 86.0		An 7.7	Hc 1.1
	Wo 54.6			Sp 33.4	Or 0.7		Or 0.4	Sp 6.4
				Uv 2.8			Cs 90.2	Uv 65.1
				Mt 3.6				Mt 25.6

* Fe_2O_3 calculated to achieve stoichiometry.
≠ Ni below detection limit.
+ Includes 0.01% Y_2O_3 and 0.25% Ce_2O_3. Analysis by A.A. Chodos, CalTech.
n.d. = not determined.

spinel occurs as minute grains of dark green color distributed throughout the meteorite. It is homogeneous and rich in Fe and Al (i.e., hercynitic spinel; Table 9; see also Hutchison, 1972). Celsian is found as small grains in the groundmass. The grains are compositionally homogeneous and contain about 90% of the celsian molecule (Table 9). Angra dos Reis is the only meteorite having magnesian kirschsteinite and celsian. A homogeneous and highly calcic plagioclase, $An_{86.0}$ (Table 9), was found only in mineral separates. Whitlockite (merrillite) is inhomogeneously distributed, quite variable in size and may represent about 0.3% of the meteorite (Ludwig and Tschermak, 1887, 1909). X-ray refinement of its structure (Dowty, 1977) indicates that the phase is better referred to as merrillite. Magnetite forms small and rare grains containing 21.9% TiO_2 (Table 9), whereas metallic nickel-iron (Fe 95.5, Ni 3.5, Co 1.15, total 100.15%) is very rare and mainly occurs as tiny blebs in troilite (Fe 62.9, Co 0.07, S 36.5, total 99.47%; cf., Keil et al., 1976). On the basis of their bulk chemical analysis, Ludwig and Tschermak (1887, 1909) calculated 1.26% troilite for the meteorite, and this seems to be a reasonable value. Minute grains of baddeleyite were recognized with the electron microprobe on the basis of their high Zr contents but were too small for quantitative analysis. Terrestrial hydrous iron oxide, apparently the result of weathering of troilite grains, is found as a rare constituent.

Chemical composition — The bulk chemical analysis of Angra dos Reis (Table 10) shows as main features high TiO_2 and CaO contents markedly different from any other meteorite. Additional major element data are in Smales et al. (1970) and Ma et al. (1977). Trace element abundances in clinopyroxene, olivine and whitlockite separates are listed in Table 12. With thc exception of a few elements, abundances for the whole rock (Table 13) and the fassaite are very similar, reflecting the fact that Angra dos Reis is essentially a pyroxenitic rock. The meteorite shows a small negative Eu anomaly (Fig. 22), interpreted as due to the crystallization and removal of plagioclase (Schnetzler and Philpotts, 1969) or as the result of early melilite crystallization, a phase that is speculated to have been removed in the cumulate process (Prinz et al., 1977b).

Oxygen isotopic compositions of pyroxene ($\delta^{18}O = 3.66$, 3.86; $\delta^{17}O = 1.75$, 1.73) and olivine ($\delta^{18}O = 3.12$; $\delta^{17}O = 2.00$) in Angra dos Reis have been determined by Clayton et al. (1976). The points fall close to the terrestrial fractionation line. However, the results apparently imply disequilibrium conditions between olivine and pyroxene as indicated by the non-paralled trend of its tie-line relative to the terrestrial one.

Nuclear particle tracks — Bhandari et al. (1971a, b) determined a mean fission track density of 12×10^6 cm^{-2} (total range: $10.5 - 18.5 \times 10^6$ cm^{-2}) in Angra dos Reis pyroxenes. On the basis

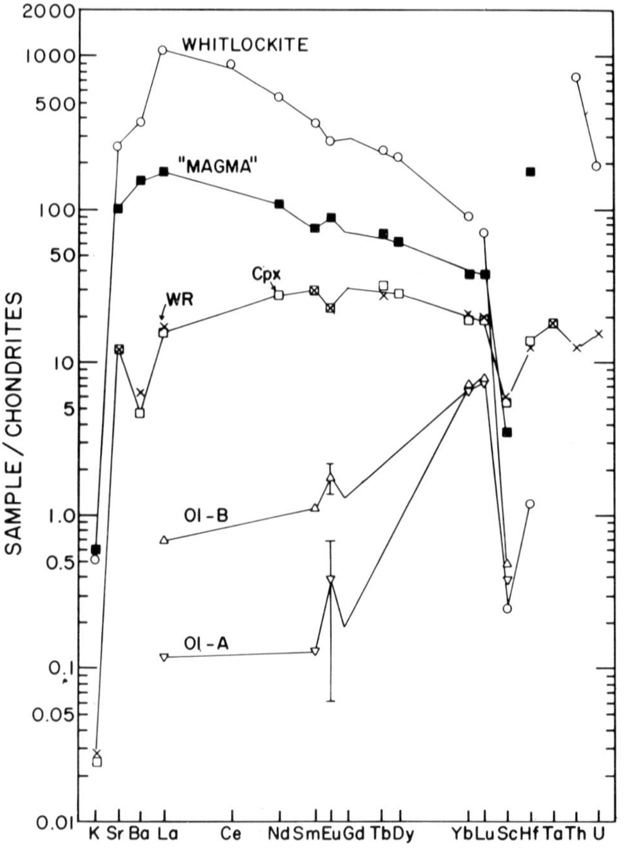

Figure 22 — Trace element abundances in the Angra dos Reis, Rio de Janeiro, Brazil fassaite achondrite (Ma et al., 1977). WR stands for whole rock; Cpx, clinopyroxene; 01-A and 01-B, olivine A and olivine B, respectively; "magma", a calculated differentiated magma from which the Angra dos Reis clinopyroxene could have crystallized (Reprinted with permission of Elsevier Publ. Comp.)

of track-length measurements, they estimated a contribution of cosmic ray tracks of less than 10^4 cm^{-2}. More recently Carver and Anders (1976a) reported a mean fission track density of $7.8 \pm 0.8 \times 10^6$ cm^{-2}, and a cosmic ray track component of $1.3 - 4.4 \times 10^6$ cm^{-2}. Furthermore, Störzer and Pellas (1976, 1977) obtained cosmic ray track densities of 1.3×10^6 cm^{-2} in feldspars, olivines and pyroxenes, leading to an estimated pre-atmospheric mass of Angra dos Reis of about 80 kg. According to the same authors, the ^{244}Pu fission track densities in pyroxenes and whitlockites are about 20 and 280×10^6 cm^{-2}, respectively, which correspond exactly to the ^{244}Pu fission xenon contents in the two minerals. This suggests that the Angra dos Reis rock cooled at a rate faster than 70° K per m.y. between the temperature interval 1040 - 320° K.

Table 10. Bulk-chemical analyses of Brazilian stone meteorites

	SiO_2	TiO_2	Al_2O_3	Cr_2O_3	Fe_2O_3	FeO	MnO	MgO	CaO	Na_2O	K_2O
Angra dos Reis	43.94	2.39	8.73	tr	0.31	8.28		10.05	24.51	0.036*	0.0049*
Governador Valadares	49.52	0.35	1.74	0.21	1.14	18.62	0.67	10.92	15.82	0.82	0.43
Ibitira	49.0	0.78	12.6	0.36		18.1	0.50	7.13	10.9	0.19	0.01
	48.81	0.90	12.97	0.38		18.19	0.47	7.28	10.99	0.21	0.03
Serra de Magé	43.42		27.20	0.33	2.07	4.79	0.58	3.18	14.53	1.59	0.20
	48.42	0.166	12.69	0.54		16.18	0.550	11.35	9.08		
	46.69	0.11	20.89	0.33		9.97	0.36	7.52	13.09	0.30	0.00
Avanhandava	36.70	0.05	3.50	tr		10.21	0.24	24.80	2.32	0.77	0.20
	36.65	0.12	2.10	0.48		9.95	0.33	23.21	1.68	0.78	0.09
Conquista	35.99	0.11	2.07	0.47	6.00	8.04	0.31	22.52	1.73	0.81	0.09
Ipiranga	36.58	0.26	2.94	0.10		12.68	0.50	23.61	1.36	0.84	0.16
	37.10	0.12	2.17	0.52		9.90	0.27	23.55	1.77	0.90	0.11
Itapicuru Mirim	35.61	0.11	2.05	0.53		9.44	0.32	22.31	1.63	0.79	0.09
Macau	36.26	0.11	2.02	0.52	6.19	8.83	0.32	22.74	1.65	0.84	0.09
Mafra+	38.14	0.080	2.07	0.475		10.74	0.287	23.29	1.81	0.91	0.092
Marília	38.38	0.073	2.19	0.484		8.54	0.294	23.18	2.00	1.01	0.095
	37.09	0.11	2.06	0.49		9.82	0.32	23.26	1.66	0.84	0.09
São José do Rio Preto	36.70	0.11	2.49	0.54		10.02	0.31	23.08	1.83	0.86	0.09
Uberaba	37.50	0.12	2.30	0.55		11.52	0.32	23.43	1.78	0.90	0.09
Minas Gerais	39.32	0.12	2.41	0.60		15.27	0.35	24.83	2.00	0.96	0.11
Paranaíba	40.22	0.14	2.17	0.45		13.37	0.37	24.96	1.92	0.96	0.11
Patrimônio	39.84	0.11	2.06	0.47		16.11	0.33	24.36	1.84	0.91	0.10
Putinga	40.4	0.1	2.1	0.5		16.3	0.3	24.4	1.8	1.0	0.1
	38.95	0.13	2.35	0.52		17.03	0.34	24.32	1.80	0.95	0.11
Rio Negro	39.32	0.12	2.18	0.56		14.54	0.34	24.26	1.77	0.90	0.10
Parambu	40.80	0.092	2.12	0.531		18.77	0.332	24.31	2.03	1.02	0.109

† High metallic iron due to the presence of large metal inclusion (0.5g) in 12g sample.

≠ Excessive C content is thought to be largely due to terrestrial contamination (cutting oil, protective organically-based coatings, etc.).

ϵ Analyzed sample contains approximately 1.7% C. Two different "clean" fragments contained 0.27 and 0.36% C, respectively.

§ Calculated by converting Fe_3O_3 to Fe°, on an H_2O- and C-free basis.

+ Our data on mineral compositions (see text) clearly indicate that Mafra is an L- and not H-group chondrite.

* After Ma et al. (1977).

Table 10. Bulk-chemical analyses of Brazilian stone meteorites (cont.)

	P_2O_5	H_2O^+	H_2O^-	Fe	Ni	Co	FeS	C	Others	Total	Total Fe	References
Angra dos Reis	0.13						1.26			99.64	7.35	Ludwig and Tschermak (1909)
Governador Valadares									NiO = 0.01	100.25	14.87	Burragato et al. (1975)
Ibitira	0.05									99.62	14.1	Wänke et al. (1974)
	0.10	n.d.	0.04	n.d.	0.01	<0.01		<0.02	S = <0.1	100.38	14.14	Jarosewich (unpublished)
Serra de Magé		0.01	1.71						S = 0.04	99.83	4.44	Moraes and Guimarães (1926)
	0.057									99.03	12.58	McCarthy et al. (1973)
	0.05	0.10	0.03	n.d.	0 11	0.01		0.05	S=0.00	99.61	7.77	Prinz et al. (in preparation)
Avanhandava	0.30			15.15	1.73		6.11			102.08	27.3	Arid and Rüegg (1969)
	0.26	0.38	0.09	15.96	1.99	0.08	5.43	0.04		99.62	27.15	Paar et al. (1976)
Conquista	0.30	2.15	0.35	11.57	1.64	0.04	4.44	1.7[ε]		98.63	25.83[§]	Keil et al. (1978a)
Ipiranga	0.36	0.32	0.36	14.51	1.75	0.09	3.78	0.08		100.29	26.78	Cavarretta et al. (1975)
	0.26	0.51	0.15	14.96	1.68	0.09	5.68	0.11		99.85	26.27	Gomes et al. (1978c)
Itapicuru Mirim	0.24	0.90	0.11	19.09[†]	1.93	0.05	4.96	0.34[≠]		100.16	29.58	Gomes et al. (1977b)
Macau	0.26	1.42	0.10	12.07	1.68	0.06	4.74	0.12		100.02	26.27	Gomes et al. (1977c)
Mafra[*]	0.19			14.68	1.67	0.128	5.78			100.34	26.70	Levi-Donati et al. (1976)
Marília	0.15			16.14	1.66	0.130	5.32			99.65	26.16	Shima et al. (1974)
	0.27	0.30	0.14	16.88	1.71	0.04	5.10	0.15		100.33	27.75	Jarosewich (unpublished)
São José do Rio Preto	0.25	0.92	0.34	15.89	1.69	0.06	4.88	0.48[≠]		100.06	26.78	Gomes et al. (1978d)
Uberaba	0.25	0.57	0.05	13.50	1.73	0.05	5.70	0.09		100.45	26.08	Gomes et al. (1977a)
Minas Gerais	0.29	0.37	0.25	7.08	1.34	0.05	5.29	0.23[≠]		100.64	22.31	Jarosewich (unpublished)
Paranaíba	0.23	0.49	0.06	6.57	1.30	0.05	6.17	0.04		99.58	20.86	Keil et al. (1977)
Patrimônio	0.21	0.80	0.21	5.07	1.14	0.04	6.17	0.28[≠]		99.79	21.51	Gomes et al. (1978a)
Putinga	0.2			6.6	1.2	0.06	5.4			100.5	23.3	Symes and Hutchison (1970)
	0.27	0.81	0.06	6.03	1.15	0.07	4.96	0.20		100.05	22.42	Keil et al. (1978b)
Rio Negro	0.26	2.56	0.04	6.67	1.17	0.07	5.24	1.22[≠]		100.10	21.40	Jarosewich (unpublished)
Parambu	0.14			0.48	0.87	0.068	6.11			97.78	18.95	Shima et al. (1974)

Table 11. CIPW norms of Brazilian stone meteorites

	Fo	Fa	Ol_T	En	Fs	Hy_T	En	Fs	Wo	Di_T
Governador Valadares	6.7	9.3	16.0	3.1	3.8	6.9	14.5	18.1	32.7	65.3
Ibitira	0.0	0.0	0.0	14.7	26.2	40.9	3.2	5.7	8.7	17.6
	0.0	0.0	0.0	15.1	27.0	42.1	3.0	5.3	8.1	16.4
Serra de Magé	4.8	5.3	10.1	0.0	0.0	0.0	1.2	1.2	2.4	4.8
	0.0	0.0	0.0	26.5	28.1	54.6	2.0	2.1	4.2	8.3
	0.9	1.0	1.9	15.7	15.5	31.2	1.9	1.8	3.8	7.5
Avanhandava	32.4	11.1	43.5	13.1	4.0	17.1	1.2	0.4	1.7	3.3
	24.8	8.6	33.4	21.5	6.8	28.3	1.4	0.4	2.0	3.8
Conquista	23.1	6.7	29.8	25.0	6.6	31.6	1.6	0.4	2.2	4.2
Ipiranga	30.0	13.3	43.3	16.1	6.5	22.6	0.2	0.1	0.3	0.6
	26.4	0.9	36.3	19.9	6.1	26.0	1.6	0.5	2.4	4.5
Itapicuru Mirim	23.4	8.0	31.4	21.3	6.6	27.9	1.4	0.4	2.0	3.8
Macau	23.8	7.5	31.3	23.2	6.6	29.8	1.5	0.4	2.2	4.1
Mafra	24.4	9.1	33.5	21.2	7.2	28.4	1.8	0.6	2.7	5.1
Marília	22.8	6.8	29.6	23.0	6.2	29.2	2.3	0.6	3.3	6.2
	24.2	8.3	32.5	22.1	6.9	29.0	1.4	0.4	2.0	3.8
São José do Rio Preto	25.9	9.1	35.0	19.8	6.3	26.1	1.4	0.4	2.0	3.8
Uberaba	26.8	10.7	37.5	18.7	6.8	25.5	1.5	0.5	2.2	4.2
Minas Gerais	32.4	16.1	48.5	14.0	6.3	20.3	1.6	0.7	2.5	4.8
Paranaíba	29.5	12.9	42.4	18.9	7.5	26.4	1.9	0.7	2.8	5.4
Patrimônio	30.4	16.3	46.7	16.3	8.0	24.3	1.7	0.8	2.8	5.3
Putinga	29.9	16.2	46.1	16.1	7.9	24.0	1.7	0.9	2.8	5.4
	33.0	18.7	51.7	12.7	6.6	19.3	1.4	0.7	2.3	4.4
Rio Negro	29.6	14.4	44.0	18.2	8.0	26.2	1.5	0.7	2.3	4.5
Parambu	32.8	20.6	53.4	12.9	7.4	20.3	2.1	1.2	3.5	6.8

Note: not listed is Angra dos Reis due to the unusual pyroxene composition of that meteorite.

Table 11. CIPW norms of Brazilian stone meteorites (cont.)

	Ab	An	Or	Pl_T	Ap	Cr	Il	NiFe	Tr	Mt	Qz	Ne
Governador Valadares	6.9	0.0	2.5	9.4	0.0	0.3	0.7	0.0	0.0	1.5	0.0	0.0
Ibitira	1.7	33.8	0.0	35.5	0.0	0.6	1.5	0.0	0.0	0.0	4.0	0.0
	1.8	34.2	0.2	36.2	0.2	0.6	1.7	0.0	0.0	0.0	2.8	0.0
Serra de Magé	10.9	67.9	1.2	80.0	0.0	0.5	0.0	0.0	0.0	3.1	0.0	1.5
	0.0	35.0	0.0	35.0	0.1	0.8	0.3	0.0	0.0	0.0	0.7	0.0
	2.6	56.0	0.0	58.6	0.1	0.5	0.2	0.1	0.0	0.0	0.0	0.0
Avanhandava	6.4	5.4	1.2	13.0	0.7	0.0	0.1	16.5	6.0	0.0	0.0	0.0
	6.7	2.0	0.5	9.2	0.6	0.7	0.2	18.2	5.5	0.0	0.0	0.0
Conquista	7.3	1.9	0.6	9.8	0.7	0.7	0.2	18.5	4.7	0.0	0.0	0.0
Ipiranga	7.1	3.8	0.9	11.8	0.8	0.1	0.5	16.4	3.8	0.0	0.0	0.0
	7.7	0.7	1.6	10.0	0.6	0.8	0.2	16.9	5.7	0.0	0.0	0.0
Itapicuru Mirim	6.7	1.8	0.5	9.0	0.6	0.8	0.2	21.3	5.0	0.0	0.0	0.0
Macau	7.4	1.5	0.6	9.5	0.6	0.8	0.2	18.8	4.9	0.0	0.0	0.0
Mafra	7.7	1.3	0.5	9.5	0.4	0.7	0.2	16.4	5.8	0.0	0.0	0.0
Marília	8.6	1.2	0.6	10.4	0.3	0.7	0.1	18.0	5.3	0.0	0.0	0.0
	7.1	1.6	0.5	9.2	0.6	0.7	0.2	18.7	5.1	0.0	0.0	0.0
São José do Rio Preto	7.4	2.7	0.5	10.6	0.6	0.8	0.2	17.9	4.9	0.0	0.0	0.0
Uberaba	7.6	2.0	0.5	10.1	0.6	0.8	0.2	15.3	5.7	0.0	0.0	0.0
Minas Gerais	8.1	1.9	0.6	10.6	0.7	0.9	0.2	8.5	5.3	0.0	0.0	0.0
Paranaíba	8.2	1.3	0.7	10.2	0.5	0.7	0.3	8.0	6.2	0.0	0.0	0.0
Patrimônio	7.8	1.3	0.6	9.7	0.5	0.7	0.2	6.3	6.2	0.0	0.0	0.0
Putinga	8.4	0.9	0.6	9.9	0.5	0.7	0.2	7.8	5.4	0.0	0.0	0.0
	8.1	1.8	0.7	10.6	0.6	0.8	0.2	7.3	5.0	0.0	0.0	0.0
Rio Negro	7.8	1.7	0.6	10.1	0.6	0.8	0.2	8.1	5.4	0.0	0.0	0.0
Parambu	8.8	0.9	0.7	10.4	0.3	0.8	0.2	1.5	6.2	0.0	0.0	0.0

Table 12. Trace element abundances (ppm) in mineral separates from the Angra dos Reis fassaite achondrite

	Fassaite			Olivine		Whitlockite	
	(1)		(2)	(1)		(1)	(2)
Sc	47	45		3.0	3.9	2.0	
V	187	189		<5	9±3		
Co	10.3	8.3		31	29	3.5	
Sr			132				2790
Ba			17.4			1400	2430
La	5.8	5.4		0.041	0.23	370	
Ce						750	
Nd	36±5	20±5	17.2	<3	<6	370 ± 10	389
Sm	6.1	5.9		0.025	0.21	72	84
Eu	1.74	1.66		(0.027 ±0.023)	(0.13 ±0.03)	20.5	
Tb	1.52	1.48				11.5	
Dy	8.7	8.5				66	
Yb	4.5	4.2		1.40	1.51	20	
Lu	0.68	0.65		0.25	0.28	2.4	
Hf	3.1	2.8				0.24 ± 0.18	
Ta	0.37	0.36					
Th						29	30.3
U				<0.2	<0.2	1.6 ± 1.1	1.79

References:
(1) Ma et al., 1977; (2) Wasserburg et al., 1977.

Age — U-Pb, Th-Pb and Pb-Pb ages for the total meteorite (Table 2) and for high-purity whitlockite are concordant at 4.54 b.y. (Wasserburg et al., 1977). A pyroxene-phosphate internal isochron age of 4.55 ± 0.04 b.y. (Table 2, Fig. 7) by Lugmair and Marti (1977) and a $^{207}Pb/^{206}Pb$ age of 4.555 ± 0.005 b.y. by Tatsumoto et al. (1973) agree well with those data. However, older (U-He 5.0 ± 0.3 b.y., Müller and Zähringer, 1969; U-He 5.1 b.y., Ganapathy and Anders, 1969) and younger (Rb-Sr 1.3 b.y., Birck et al., 1975; K-Ar < 1.93 b.y., Müller and Zähringer, 1969; K-Ar 1.6 b.y., Ganapathy and Anders, 1969) ages were also reported in the past, and are clearly erroneous or do not represent crystallization ages. Carver and Anders (1976a) have suggested to express the paradoxical high values not as ages but as excesses of ^{4}He.

The Angra dos Reis meteorite also has a primitive $^{87}Sr/^{86}Sr$ ratio (Papanastassiou and Wasserburg, 1969; Papanastassiou, 1970). It is distinctly lower than BABI and nearly as low as those in some refractory Allende inclusions which are thought to be among the earliest condensates of solar system material (Wasserburg et al., 1977). By using the data from the whole rock and from pyroxene and

Table 13. Trace element abundances in the Angra dos Reis fassaite achondrite

	(1)	(2)	(3)	(4)	(5)	(6)	(7)	(8)		(9)	(10)	
Sc, ppm					57						46	46
V, ppm					160(150)						188	189
Cr, ppm					1700							
Mn, ppm					700							
Co, ppm					~10		20.4				11.4	13.3
Ni, ppm					30(50)							
Cu, ppm					10							
Zn, ppm					10		2.1					
Ga, ppm					6		0.36					
Br, ppb		790					413					
Rb, ppm					1					0.063		
Sr, ppm					100(150)					135		
Y, ppm					40(30)							
Zr, ppm					100(100)							
Nb, ppm					5							
Ag, ppb							18					
In, ppb							1.43					
Cs, ppb			0.44				1.9					
Ba, ppm				26.4	30(30)					26.5		
La, ppm					10						5.7	5.9
Ce, ppm				20.6	20							
Pr, ppm					10							
Nd, ppm				19.8	20						16 ± 3	19 ± 3
Sm, ppm				6.45	8						5.9	6.0
Eu, ppm				1.81	2.0						1.64	1.69
Gd, ppm				8.4	10							
Tb, ppm					2						1.31	1.29
Dy, ppm				9.11	12						8.6	9.2
Ho, ppm					3.5							
Er, ppm				5.31	5							
Tm, ppm					1							
Yb, ppm				4.73	6						4.7	4.8
Lu, ppm					0.7						0.69	0.73
Hf, ppm					3						2.6	2.5
Ta, ppm											0.38	0.35
Ir, ppb						⩾2	2.6					
Au, ppb						1.9	7.17					
Tl, ppb							0.86					
Pb, ppm								0.541	0.550			
Bi, ppb							2.4					
Th, ppm	0.967				1			0.870	0.901	0.51		
U, ppm	0.198				0.4			0.203	0.211	0.16	0.31 (±0.11)	<0.2

References:
(1) Morgan and Lovering, 1964; (2) Liebermann and Ehmann, 1967; (3) Tera et al., 1970; (4) Schnetzler and Philpotts, 1969; (5) Smales et al., 1970; in parentheses, data obtained using different methods; (6) Ehmann et al., 1970; (7) Laul et al., 1972; (8) Tatsumoto et al., 1973; (9) Wasserburg et al., 1977; (10) Ma et al., 1977.

whitlockite separates, Wasserburg et al. (1977) give a value of 0.69883 ± 0.00002 as the preferred one for the initial $^{87}Sr/^{86}Sr$. This figure is identical to the value of 0.69884 ± 0.00004 reported by Papanastassiou (1970).

An exposure age, as obtained by the $^{81}Kr/^{83}Kr$ method, of 55.5 ± 1.2 m.y. was reported by Lugmair and Marti (1977); other values listed in the literature are 51 ± 5 m.y. (Carver and Anders, 1976a), 58 m.y. (Müller and Zähringer, 1969), 68 m.y. (Munk, 1967; Ganapathy and Anders, 1969) and 71.5 ± 2.5 m.y. (Birck et al., 1977). Cosmic-ray produced ^{26}Al in Angra dos Reis has been measured by Heymann and Anders (1967) and Fuse and Anders (1969), the results being slightly different (113 ± 9 and

97.2 ± 4.7 dpm/kg, respectively), but still within the statistical uncertainties.

Fission xenon anomalies in Angra dos Reis were first noted by Munk (1967) and, since then, investigations regarding the isotopic composition of Xe in the meteorite have been reported by many authors (Müller and Zähringer, 1969; Hohenberg, 1970; Podosek and Huneke, 1971; Podosek, 1972; Lugmair and Marti, 1977; Wasserburg et al., 1977).

Origin — Due to its unusual features — a very primitive $^{87}Sr/^{86}Sr$ ratio, old age of 4.55 b.y., evidence for extinct ^{244}Pu, etc. — the Angra dos Reis fassaite achondrite is an igneous rock of remarkable and complex history. Extensive isotopic investigations of the meteorite by Wasserburg et al. (1977) establish Angra dos Reis as an early planetary differentiate which has not been disturbed since 4.5 b.y. ago. Thus, Angra dos Reis, which is clearly a cumulate igneous rock that formed by a series of complex processes, is as old as the solid matter in our solar system, within the error of the age measurements. Its content of the decay products of extinct radionuclides suggests that the time interval between nucleosynthesis and formation of its parent body was also very short, possibly less than 50 m.y. Thus, it appears that the time period from nucleosynthesis through condensation of solids, accumulation of the solids to form the parent body of Angra dos Reis, melting or partial melting of that parent body and formation of the fassaite rock through complex partial melting-cumulate processes, as well as cooling to temperatures sufficiently low to retain tracks and noble gases, was an incredibly short 50 m.y. or less.

Storzer and Pellas (1977) emphasized that there is no indication that after its rapid cooling at 4.55 b.y., Angra dos Reis was again exposed to higher temperatures. On the basis of textural and mineralogical data, Prinz et al. (1977b) concluded that the meteorite crystallized at low pressures from a highly SiO_2 undersaturated melt. Since it formed as a cumulate, the pyroxene crystals gravitationally settled from a melt which may have crystallized melilite first. Thus, the negative Eu anomaly of the meteorite can be explained by the early crystallization of melilite, which was removed in the cumulate process. Measurements of the oxygen fugacity as a function of temperature made by Brett et al. (1977) on an interior bulk sample of the meteorite are compatible with crystallization and cooling at pressures greater than 50 bars. The chemical study of the whole rock and mineral separates olivine, clinopyroxene and whitlockite allowed Ma et al. (1977) to derive some conclusions on the genesis of the meteorite. They propose for its origin a two-stage evolutionary model, including an early partial melting from a source material composed of olivine, orthopyroxene and clinopyroxene, followed by a later crystallization.

Available material — 1) Museu Nacional, Rio de Janeiro, Brazil (4 sp, 101.8 g); 2) Muséum National d'Histoire Naturelle, Paris, France (2 sp, 15 g); 3) Naturhistorisches Museum Wien, Vienna, Austria (2 sp, 7 g); 4) United States National Museum (Smithsonian Institution), Washington, D.C. (1 sp, 6.5 g); 5) Field Museum of Natural History, Chicago, U.S.A. (1 sp, 6.4 g); 6) British Museum (Natural History), London, England (1 sp + probe section, 5.2 g); 7) Humboldt University, Berlin, German Democratic Republic (2 g, cf., Hutchison et al., 1977); 8) Institute of Meteoritics, Albuquerque, U.S.A. (probe sections).

7.1.2. Augite-olivine achondrite

Governador Valadares (18°51'S; 41°57'W)

Fall circumstances — The Governador Valadares, State of Minas Gerais, meteorite was found in 1958 by a mineral hunter (garimpeiro), near the city of Governador Valadares. Unfortunately, details on time and exact coordinates of the fall are unknown. One single specimen weighing 158 g is said to have been found. The stone is shiny and well-preserved in shape and almost completely covered by a black, glassy fusion crust. On the basis of the mineralogy and chemical composition, Burragato et al. (1975) classified the meteorite as an augite-olivine achondrite (nakhlite), only the third known specimen of this meteorite class.

Texture — The Governador Valadares meteorite shows a very conspicuous porphyritic texture (Fig. 23). Augite phenocrysts are embedded into a fine-grained interstitial material consisting of glass and semi-radiating skeletal crystals of plagioclase, K-feldspar, silica, chlorapatite, pigeonite and opaque phases. Textural and compositional data suggest that most or all olivine is also an interstitial phase. The overall texture can be properly definied as trachytic-porphyritic with variolitic mesostasis. Augite grains, commonly exhibiting wide twin bands (Fig. 24), have elongated tabular habits, which — as confirmed by universal stage measurements — are preferentially aligned to form a mineral lineation (Berkley et al., 1979). The textural evidence suggests that the rock represents a cumulate which was in the act of flow during or after crystal settling. Slight evidence for shock metamorphism is indicated by the wavy extinction of silicate grains and chevron-like lamelar augite glide planes displaying kinking as well.

Mineralogy — According to Berkley et al. (1979), the meteorite consists essentially of subhedral to euhedral augite with minor amounts of anhedral to subhedral interstitial olivine. Augite is weakly zoned, with Fe/Mg slightly increasing toward grain margins;

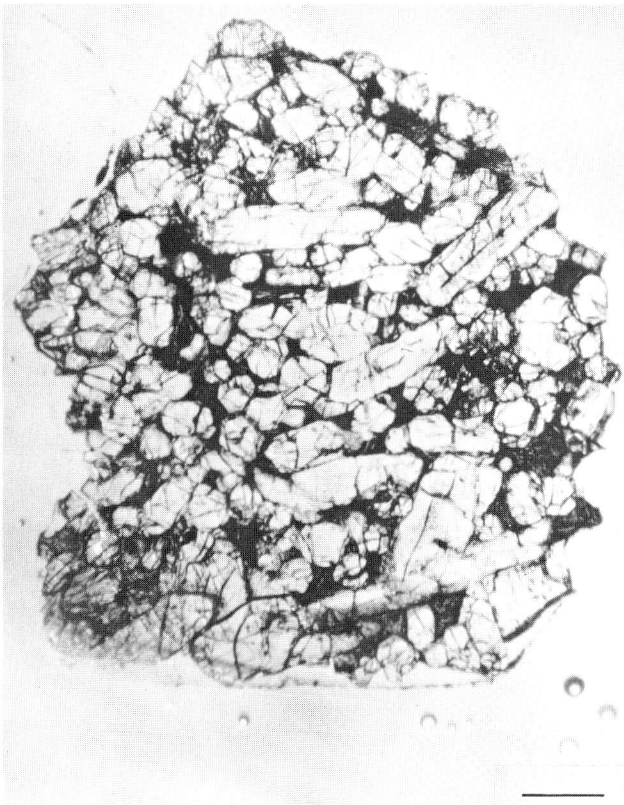

Figure 23 – Photomicrograph of the Governador Valadares, Minas Gerais, Brazil augite-olivine achondrite. Most grains are augite, with the prismatic ones displaying a preferred orientation. Dark is fine-grained, interstitial material. Transmitted, plane polarized light. Scale bar equals 0.61 mm.

average composition is $En_{38.9}Fs_{22.6}Wo_{38.5}$ (Table 14). Olivine shows more pronounced compositional zoning, in some cases varying by as much as 7.0 mole % Fa (core, Fa_{63}; rim, Fa_{70}). Core compositions also vary from grain to grain (Fa_{68-62}), the average being $Fa_{66.2}$ (Table 14). Olivine is found frequently veined or mantled by a brownish-red alteration product identified as iddingsite. Minor constituents of the meteorite include interstitial pigeonite (compositions ranging from $En_{32-21}Fs_{61-66}Wo_{7-13}$; average composition is $En_{26.7}Fs_{63.9}Wo_{9.4}$), plagioclase ($Ab_{48.3}An_{48.8}Or_{2.9}$), alkali feldspar ($Ab_{31.0}An_{5.0}Or_{64.0}$), titanomagnetite ($Cm_{2.9}Uv_{44.4}Hc_{0.6}Sp_{2.6}Mt_{49.5}$), rare ilmenite (Table 14), and troilite, chalcopyrite, pyrite, chlorapatite and pure SiO_2.

Chemical composition – Bulk chemical data on Governador Valadares (Table 10) agree well with those listed in the literature for the Lafayette and Nakhla augite-olivine achondrites. The

Figure 24 — Photomicrograph of the Governador Valadares, Minas Gerais, augite-olivine achondrite, showing abundant twinning in augite. Transmitted light, crossed nicols. Scale bar equals 0.23 mm.

pyroxenitic composition of the meteorite is quite apparent (Table 11), with normative hypersthene and diopside making up 72% of the rock; additional minerals include olivine and plagioclase.

Age — The $^{40}Ar/^{39}Ar$ age of Governador Valadares has been determined by Bogard and Husain (1977) to be 1.32 ± 0.04 b.y., which is essentially identical to the ages for the other known augite-olivine achondrites previously reported by Podosek (1973). This fact strongly indicates that all three meteorites were derived from the same parent body and, possibly, the same rock unit. A recent Rb-Sr study of Governador Valadares by Wooden et al. (1979) also gives a low age, 1.33 ± 0.01 b.y., which has been assumed to represent the age of crystallization for the augite-olivine achondrites. Bogard and Husain (1977) also list a cosmic ray exposure age of 8 ± 1 m.y. for the meteorite.

Origin — Berkley et al. (1979) suggest that Governador Valadares is a cumulate rock that formed by relatively slow cooling during augite accumulation, followed by the quenching of the intercumulus mesostasis. Sudden eruption to the surface or near surface of a deeper-formed crystal-liquid assemblage could be a possible mechanism to explain the two-stage cooling history indicated by the texture of the rock. Crystallization took place under relatively highly

Table 14. Average compositions of silicate and oxide phases in the Governador Valadares augite-olivine achondrite

	Cumulate augite (1)	Interstitial pigeonite (1)	Olivine (1)	Plagioclase (2)	Alkali feldspar (2)	Titano-magnetite (2)	Ilmenite (2)
SiO_2	51.3	48.3	34.0	58.8	73.0	n.d.	n.d.
TiO_2	0.17	0.42	n.d.	n.d.	0.16	15.7	45.0
Al_2O_3	0.81	0.35	n.d.	24.6	14.1	1.43	0.72
Cr_2O_3	0.36	0.11	<0.09	n.d.	n.d.	1.27	0.90
V_2O_3	n.d.	n.d.	n.d.	n.d.	n.d.	0.63	0.42
Fe_2O_3						34.8*	15.6*
FeO	14.0	36.6	49.9	0.88	n.d.	44.2	36.1
MnO	0.37	0.95	0.71	n.d.	n.d.	0.60	0.88
MgO	13.5	8.6	14.3	0.22	0.90	0.46	1.41
CaO	18.6	4.2	0.44	6.4	0.77	n.d.	n.d.
Na_2O	0.17	<0.04	n.d.	7.0	2.67	n.d.	n.d.
K_2O	n.d.	n.d.	n.d.	0.64	8.38	n.d.	n.d.
ZrO_2	n.d.	n.d.	n.d.	n.d.	n.d.	0.07	0.12
TOTAL	99.28	99.55	99.35	98.54	99.98	99.16	101.15
	En 38.9	26.7	Fo 33.8	Ab 48.3	31.0	Cm 2.9	
	Fs 22.6	63.9	Fa 66.2	An 48.8	5.0	Uv 44.4	
	Wo 38.5	9.4		Or 2.9	64.0	Sp 2.6	
						Hc 0.6	
						Mt 49.5	

* Calculated on the basis of stoichiometry.
References:
(1) Berkley et al., 1979; (2) Berkley, J.L., unpublished data.
n.d. = not determined.

oxidizing conditions, as is indicated by the presence of titanomagnetite and water-bearing iddingsite of extraterrestrial origin.

The three known augite-olivine achondrites, Nakhla, Lafayette and Governador Valadares, are very similar in bulk and mineral compositions as well as in age (Boctor et al., 1976; Bogard and Husain, 1977). This has prompted some to suggest that all, in fact, are pieces of one and the same meteorite (note that only Nakhla is an observed fall, whereas Lafayette and Governador Valadares are finds and, thus, could be mislabeled and misplaced specimens of Nakhla). However, recent detailed studies of all three meteorites by Berkley et al. (1979) indicate subtle but nevertheless notable differences between the three stones: 1) CaO is significantly lower in large olivine grains in Lafayette than in the other two nakhlites; 2) compositions of interstitial clinopyroxene are dramatically more enstatite-rich in Lafayette; 3) mesostasis areas in Lafayette are more coarse-grained and glass-rich; and 4) Al_2O_3 in cumulus augite differs between individual nakhlites. Thus, unless extreme heterogeneities between different specimens are the cause for the observed differences noted by Berkley et al. (1979), these authors conclude that the three rocks are not from one and the same meteorite fall. However, they clearly formed under similar conditions and at the same time and, thus, may represent samples of the same rock formation on their parent body.

The young ages of Governador Valadares and the other augite-olivine achondrites are remarkable. Textural studies indicate that these rocks have not been affected by relatively recent secondary processes such as shock reheating that could have reset the time clock and explain their young ages. On the other hand, thermal considerations make it most unlikely that the relatively small planetesimals envisaged as parent bodies for Governador Valadares and the other augite-olivine achondrites could have been heated to the melting points of silicates by internal heat sources *that* recently in the history of the solar system. Thus, one is forced to postulate external heat sources, such as complete impact melting, although there is no evidence for such a process (e.g., no residual clasts) in these rocks. Thus, the young ages of the augite-olivine achondrites and the heat source for magma generation are still a mystery.

Available material — 1) Museo di Mineralogia, Rome, Italy (1 sp, 97 g); 2) British Museum (Natural History), London, England (1 sp, 6.4 g).

7.1.3. Monomict pigeonite-plagioclase achondrites

Ibitira (19° 22'S; 45° 14'W)

Synonyms — Monjolo Farm, Martinho Campos.

Fall circumstances — The Ibitira, State of Minas Gerais, meteorite fell in the afternoon (between 5.00 and 5.30 hr) of June 30, 1957, in the county of Martinho Campos, at the Monjolo Farm near the village of Ibitira. According to witnesses, sound resembling thunder followed by whistling-like noise and light effects (fireball traveling northwestward, visible up to 160 km away from the fall place) were noticed. One single stone weighing 2,500 g was reported to have been recovered. It excavated a small pit about 25 cm deep and about 20 cm in diameter. The fall was reported by Menezes (1957) and the meteorite is listed in Hey (1966) and Hutchison et al. (1977).

Texture — The most impressive textural feature of Ibitira is the presence of vesicles (Wänke et al., 1972), nearly spherical and very uniform in size, comprising roughly 5-7% by volume of the rock (Fig. 25). The meteorite is holocrystalline, fine-grained and shows a very pronounced hornfelsic texture (Fig. 26). No evidence of brecciation was observed in the samples studied. However, mineralogical and textural evidence indicates that the meteorite was subjected to strong shock metamorphism which mainly affected plagioclase and tridymite (Steele and Smith, 1976b). The surface of the meteorite is covered by melting crust exhibiting abundant flight markings (Fig. 27).

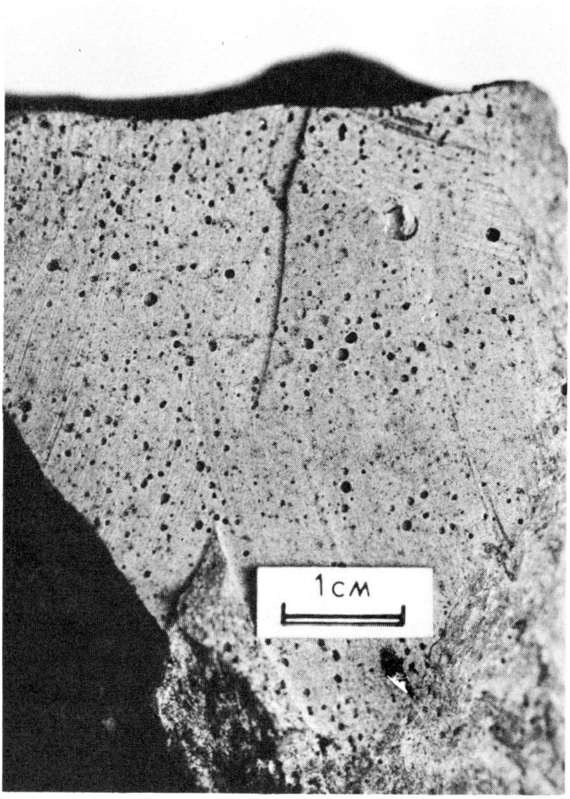

Figure 25 — Photograph of a sawed face of the Ibitira, Minas Gerais, Brazil monomict pigeonite-plagioclase achondrite, showing a large number of mm-sized vesicles.

Mineralogy — According to Wilkening and Anders (1975) and Steele and Smith (1976a, b), the meteorite consists essentially of pyroxene (60 vol. %) and plagioclase (30%), with minor amounts of tridymite (5%) and accessory amounts of olivine, chromite, ilmenite, metallic nickel-iron (kamacite), troilite and merrillite totaling 5% of the rock.

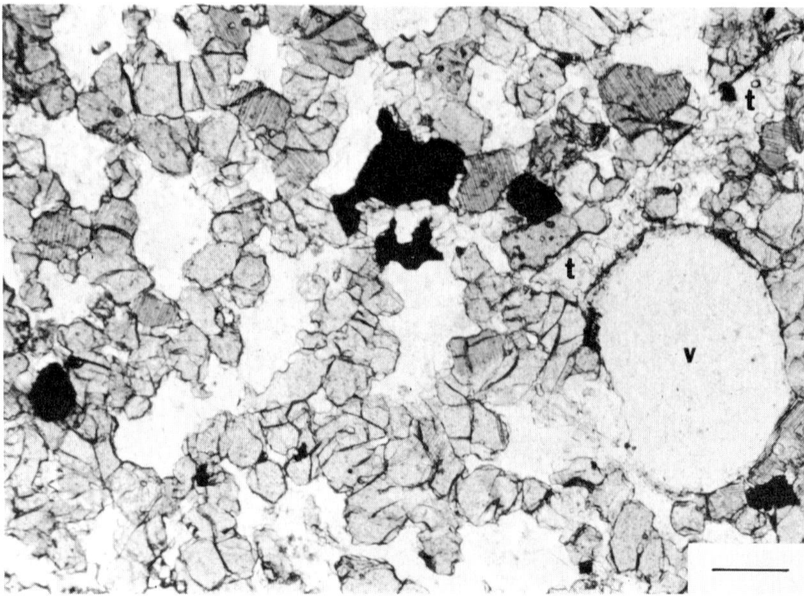

Figure 26 – Photomicrograph of the Ibitira, Minas Gerais, Brazil monomict pigeonite-plagioclase achondrite, showing plagioclase (white), equant grains of pale-brown pyroxene with exsolution lamellae, ilmenite-chromite intergrowth (black), tridymite (t) with inclusions of plagioclase, and large, round vesicles (v). Transmitted, plane polarized light. Scale bar equals 0.18 mm.

Pyroxene is pigeonite ($En_{39.0}Fs_{57.3}Wo_{3.7}$) and has well-developed exsolution lamellae of augite ($En_{31.4}Fs_{26.7}Wo_{41.9}$) in most grains (Table 15). The bulk composition of the pyroxene, as obtained by broad beam analysis, is $En_{38}Fs_{48}Wo_{14}$ (Wilkening and Anders, 1975) and $En_{37}Fs_{48}Wo_{15}$ (Steele and Smith, 1976b). This composition corresponds to a Ca-rich pigeonite as the primary pyroxene before exsolution. Plagioclase is unzoned and exhibits mosaic extinction and variable structural state. Its composition is remarkably constant, averaging An_{95} (Table 15). A similar value was obtained for a grain from a vesicle and for inclusions in tridymite. Fe-olivine ($Fa_{83.3}$) (Table 15) forms a narrow rim on one ilmenite-Ti-chromite cluster. Chromite and ilmenite are the more common opaque minerals in the meteorite. They are intergrown with one another and have uniform compositions in each intergrowth. However, the compositions are quite variable from one intergrowth to another. Most chromites have high TiO_2 (~15-25%), but in one analysis it is low (TiO_2, 5.71%) (Table 15; see also Table 2 in Wilkening and Anders, 1975). Grains picked from the vesicles have identical compositions. Nickel-iron (kamacite) is commonly associated with troilite, but troilite may also occur in isolated grains. Troilite is also found in association with chromite-ilmenite or as round droplets in pyroxene grains. Data by Steele and Smith (1976b) on the

Figure 27 — Ibitira, Minas Gerais, Brazil monomict pigeonite-plagioclase achondrite, weighing 2,500 g and showing a shiny fusion crust and abundant linear flight markings. Vesicles are visible in places with no fusion crust.

nickel-iron phase gave different Co and Ni contents, depending on the nature of the associated minerals. For example, in metal associated with troilite, Co ranges from 0.80 to 1.10% and Ni (~0.88%) is nearly constant. In metal associated with opaque phases, these concentrations are lower (Co, 0.40-0.60%; Ni, 0.10-0.35%). Tridymite occurs as plates up to 2 mm long and contains small plagioclase inclusions comprising about one-quarter of the volume. The laths are commonly associated with the vesicles. Compositional data for two grains are listed in Table 15. Merrillite is found as pale-brown euhedral grains in the vesicles (Fig. 28).

Chemical composition — Bulk chemical analyses of Ibitira are listed in Table 10 and agree well with each other. The CIPW norms (Table 11) show hypersthene, diopside and plagioclase as major minerals; quartz is a minor constituent.

Ibitira is significantly depleted in the volatile alkalis Na, K and Rb but not in Cs, relative to the other eucrites (Stolper, 1977). Trace element abundances are given in Table 16 and are plotted in

Table 15. Average compositions of selected minerals in the Ibitira monomict pigeonite-plagioclase achondrite

	Low-Ca pyroxene (1)	High-Ca pyroxene (2)	Average pyroxene (3)	Olivine (4)	Plagioclase ≠21 (5)	Plagioclase ≠27 (5)	Ilmenite (6)+	Ilmenite (7)	High-Ti chrom. (8)	Low-Ti chrom. (8)	Silica § ≠24 (9)	Silica § ≠29 (9)
SiO_2	50.7	50.1	50.3	32.1	44.35	44.78	n.d.	n.d.	n.d.	n.d.	98.65	98.71
TiO_2	0.46	1.02	0.66	0.42*	n.d.	n.d.	54.2	51.9	17.6	5.71	n.d.	n.d.
Al_2O_3	0.23	1.50	0.55	n.d.	35.85	35.89	n.d.	0.1	4.79	7.60	0.21	0.18
Cr_2O_3	n.d.	0.29	0.39	1.04*	n.d.	n.d.	n.d.	0.4	30.5	50.6	n.d.	n.d.
FeO	34.0	16.1	28.4	60.2	0.29	0.21	45.4	45.3	47.3	34.7	0.34	0.21
MnO	0.84	0.29	0.77	0.84	n.d.	n.d.	0.51	n.d.	n.d.	n.d.	n.d.	n.d.
MgO	13.0	10.6	12.2	6.76	n.d.	n.d.	1.00	0.9	0.82	1.31	n.d.	n.d.
CaO	1.71	19.7	7.12	0.13	18.57	18.65	n.d.	n.d.	n.d.	n.d.	0.04	0.05
Na_2O	n.d.	n.d.	n.d.	n.d.	0.53	0.50	n.d.	n.d.	n.d.	n.d.	0.00	0.01
K_2O	n.d.	n.d.	n.d.	n.d.	0.07	0.08	n.d.	n.d.	n.d.	n.d.	0.13	0.13
TOTAL	100.94	99.60	100.39	101.49	99.66	100.11	101.11	98.6	101.01	99.92	99.37	99.29
	En 39.0	31.4	36.7	Fo 16.7	Ab 4.9	4.6			Cm 42.1	68.6		
	Fs 57.3	26.7	47.9	Fa 83.3	An 94.7	94.9			Hc 5.4	8.0		
	Wo 3.7	41.9	15.4		Or 0.4	0.5			Uv 48.1	16.1		
									Sp 4.4	7.3		

References:
1-4, 6, 8 (after Steele and Smith, 1976a,b); 5, 7, 9 (after Wilkening and Anders, 1975).
* Values high due to secondary fluorescence of Cr-ulvöspinel.
\+ Coexisting with high-Ti chromite.
§ Tridymite according to Steele and Smith (1976a, b).
n.d. = not determined.

Table 16. Trace element abundances in the Ibitira monomict pigeonite-plagioclase achondrite

	(1)	(2)
Sc, ppm	31.5	30
Co, ppm	14.3	7.6
Ni, ppm		6
Cu, ppm	5.1	
Zn, ppm	2.7	1.6
Ga, ppm	1.02	
Ge, ppb		3.5
As, ppb	330	
Se, ppb		80
Br, ppb		70
Rb, ppb		125
Ag, ppb		14.2
Cd, ppb		30
Sb, ppb		7.3
Te, ppb		0.5
Cs, ppb		8.5
La, ppm	2.89	
Ce, ppm		10.8
Sm, ppm	1.75	2.3
Eu, ppm	0.67	0.7
Tb, ppm		0.5
Yb, ppm	1.84	1.9
Lu, ppm	0.42	0.37
Hf, ppm		1.1
W, ppm	0.14	
Re, ppt		2.0
Ir, ppt		3.9
Au, ppb	3.2	0.95
Tl, ppb		0.41
Bi, ppb		6.6
U, ppb	81	104

References:
(1) Wänke et al., 1974; (2) Higuchi and Morgan, 1975.

Fig. 29 (Wilkening and Anders, 1975) and Figs. 30 and 31 (Morgan et al., 1978). Ibitira has no Eu anomaly (Fig. 29) and a flat pattern for the refractory, incompatible elements (REE, Hf, U) (Figs. 29, 30). Except for Sc, the mean abundance of these elements is 13.1 ± 0.71 X C1 chondrites. A flat distribution is also indicated for the more volatile elements Rb, Cs, Br and Bi, whose abundance is $6.0 \pm 1.5 \times 10^{-2}$ C1. Patterns for the siderophile and chalcophile elements are not as uniform as those for the lithophiles (Fig. 31), largely being determined by silicate-metal-sulfide partitioning (Morgan et al., 1978).

Oxygen isotopic compositions were reported by Clayton and Mayeda (1975) and Clayton et al. (1976) for some minerals in Ibitira. Data for $\delta\ ^{18}O$ (plagioclase, 3.97; pyroxene, 3.48) and $\delta\ ^{17}O$ (plagioclase, 2.00; pyroxene, 1.69) indicate that the meteorite falls in the so called terrestrial group of rocks based on oxygen isotopic data; this group is made up of terrestrial and lunar samples as well as differentiated meteorites and enstatite chondrites.

Figure 28 — Merrillite from a vesicle of the Ibitira, Minas Gerais, Brazil monomict pigeonite-plagioclase achondrite. Scanning electron microscope photograph (courtesy of Ruth Fruland). Scale bar equals 33 μm.

Nuclear particle tracks — Particle tracks in Ibitira were studied by Wilkening and Anders (1975) who obtained a mean track density, mainly due to cosmic rays, of $1.8 \pm 0.3 \times 10^6$ tracks/cm^2. This suggests a lower limit of about 10 cm for the pre-atmospheric radius of the Ibitira meteorite.

Age — Light rare gases were analyzed in Ibitira by Hintenberger et al. (1965) and Heymann et al. (1968) and a cosmic ray exposure age of 10 9 m.y. (recently corrected to 8.8 m.y. by Herzog and Anders, 1971) was derived by Heymann et al. (1968).

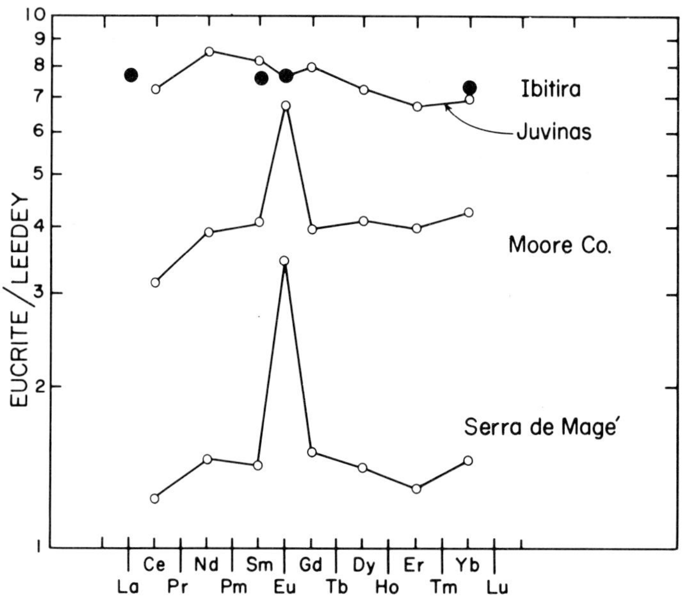

Figure 29 — Plot of the rare earth element abundances (normalized to the chondrite Leedey) for the monomict pigeonite-plagioclase achondrites Ibitira (Minas Gerais, Brazil), Juvinas (France), Moore County (U.S.A.) and Serra de Magé (Pernambuco, Brazil) (Wilkening and Anders, 1975). Except for Ibitira (Wänke et al., 1974), data are after Schnetzler and Philpotts (1969) (Reprinted with permission of Pergamon Press).

Crystallization ages were measured by Birck et al. (1975), Manhes et al. (1975) and Birck and Allègre (1978). An internal Rb-Sr isochron for the meteorite (Fig. 32) gives an age of 4.52 ± 0.25 b.y. A Pb-Pb age of 4.55 ± 0.01 b.y. was reported by Wasserburg et al. (1977). The paper by Wasserburg et al (1977) also contains some preliminary data on Xe isotopic abundances for a bulk sample of the meteorite.

Origin — Ibitira, the only known vesicular achondrite, is believed to have formed by rapid cooling of a lava flow no less than 2.5 m and probably no more than 10 or 20 m thick (Wilkening and Anders, 1975). Steele and Smith (1976b) suggested that the rock may have formed as a sub-surface dike or sill. Ibitira is one of the few unbrecciated monomict pigeonite-plagioclase achondrites (eucrites), although it shows evidence of strong shock metamorphism, giving the meteorite a recrystallized appearance. The lack of Eu anomaly (Fig. 29) (Wänke et al., 1974) suggests a liquid composition that has neither gained nor lost plagioclase. On the other hand, the distribution of refractory incompatible elements (Fig. 30) indicates that it is a relatively unfractionated rock which apparently represents a primary magma (Morgan et al., 1978).

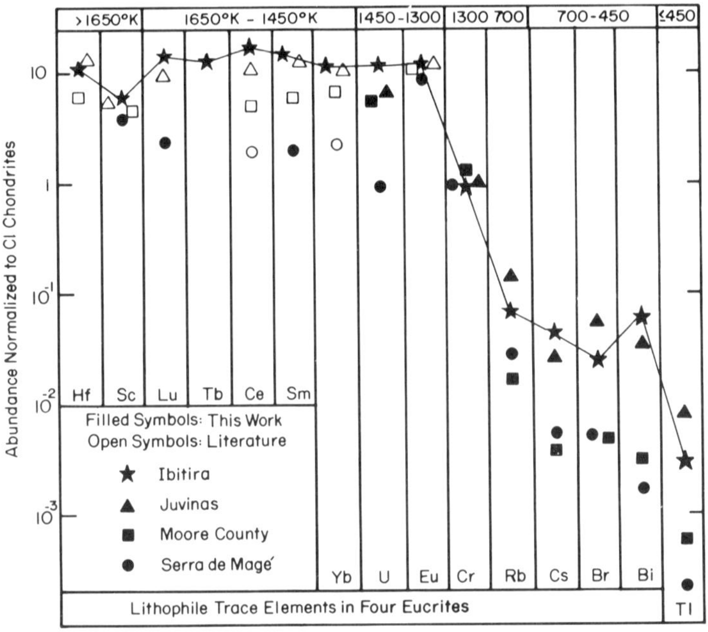

Figure 30 — Lithophile trace element abundances, normalized to C1 chondrites, in the monomict pigeonite-plagioclase achondrites Ibitira (Minas Gerais, Brazil), Juvinas (France), Moore County (U.S.A.) and Serra de Magé (Pernambuco, Brazil) (Morgan et al., 1978). Elements are arranged in approximate order of condensation temperatures (Reprinted with permission of Pergamon Press).

According to Steele and Smith (1976a, b), Ibitira has a very complex petrological history, involving at least the following processes: 1) crystallization of a basalt flow under pressure low enough to permit the formation of vesicles; 2) prolonged thermal metamorphism (hornfelsic texture, coarse exsolution of the pyroxene, and coarse intergrowth of opaque minerals), perhaps associated with vapor deposition; 3) shock metamorphism of medium intensity; 4) sufficient annealing to allow recrystallization of the plagioclase and tridymite, and partial conversion of these phases to their low structural state.

Experimental data on the petrology of eucrites (Stolper, 1977) indicate that Ibitira cannot have formed by fractionation of a liquid with a composition similar to other known monomict pigeonite-plagioclase achondrites. Yet, melt similar in composition to Ibitira could have been produced by increasing degrees of low-pressure partial melting of an olivine ($\sim Fo_{65}$), pigeonite ($\sim En_{65}Wo_5$), plagioclase ($\sim An_{94}$), Cr-rich spinel, and metal (~6-8% Ni; 0.5% Co) assemblage.

One still unsolved puzzle is the origin of the vesicles. Clearly, they are not shrinkage voids but formed by the release of gases near or at the surface of the parent body. The composition of this gas is unknown. The mineral assemblage of the rock indicates that

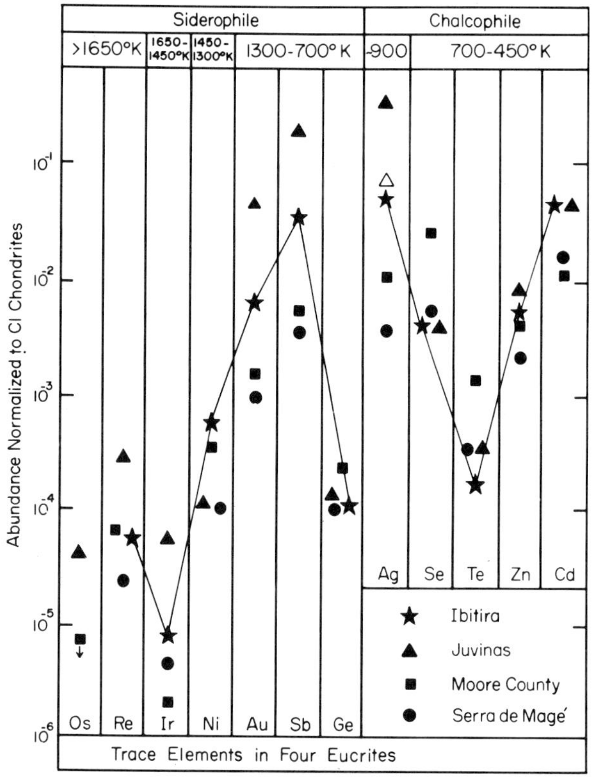

Figure 31 — Siderophile and chalcophile trace element abundances in the monomict pigeonite-plagioclase achondrites Ibitira (Minas Gerais, Brazil), Juvinas (France), Moore County (U.S.A.) and Serra de Magé (Pernambuco, Brazil) (Morgan et al., 1978). Elements are arranged in approximate order of condensation temperatures (Reprinted with permission of Pergamon Press).

the gas was not water vapor, the most common vesicle-forming gas in terrestrial rocks. We suggest that Ibitira may, in fact, have crystallized from a superheated total impact melt, with the vesicles having been formed by the release of volatile species of the common elements making up the rock (e.g., Na, K, Si, etc.). The low alkali abundances of the rock (Stolper, 1977) support this suggestion. In view of the old age of Ibitira, this implied impact melting and differentiation event must have taken place very early in the history of the solar system.

Available material — 1) Centro de Estudos Astronômicos de Minas Gerais, Belo Horizonte, Brazil (1 sp, 2,500 g; Fig. 27); 2) Instituto de Geociências, São Paulo, Brazil (1 sp, 30.7 g); 3) Institute of Meteoritics, Albuquerque, U.S.A. (1 sp, 17.9 g); 4) United States National Museum (Smithsonian Institution), Washington, D.C. (1 sp, 1.6 g); 5) Max-Planck-Institut für Chemie, Mainz, Federal Republic of Germany (powder, 2 g).

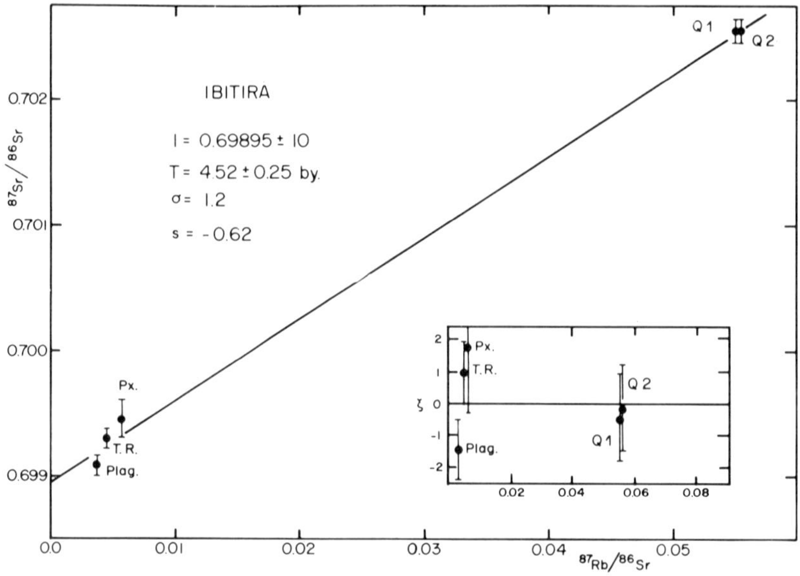

Figure 32 — Rb-Sr isotopic evolution diagram for the Ibitira, Minas Gerais, Brazil monomict pigeonite-plagioclase achondrite (Birck and Allègre, 1978). Mineral separates include pyroxene, plagioclase and low-density (d < 2.50) fractions Q1 and Q2, yielding an age of 4.52 ± 0.25 b.y. Insert illustrates error estimate (Reprinted with permission of Elsevier Publ. Comp.).

Serra de Magé (8°23'S; 36°46'W)

Synonym — Pesqueira.

Fall circumstances — The Serra de Magé, State of Pernambuco, meteorite fell in the morning (around 11.00 hr) of October 1, 1923, about 20 km from the locality of Pesqueira. After the appearance of a fireball and detonations (one loud explosion followed by sounds resembling shot gun blasts, which lasted up to 3 minutes), a shower of stones fell over an extensive area. At least some fifty fragments, most of them less than 5 cm in diameter, but some as much as 10 cm, were recovered by local people. At Garanhuns and nearby villages, windows of houses were broken and objects fell from shelves. The fall was reported by Moraes and Guimarães (1926) and the meteorite is listed in Oliveira (1931), Vidal (1936), Hey (1966) and Hutchison et al. (1977).

Texture — Moraes and Guimarães (1926), who first described the meteorite, reported the presence of light to dark green fusion crust and fine-grained, interstitial texture. Duke and Silver (1967) referred to Serra de Magé as a coarse-grained rock exhibiting equigranular to ophitic texture and gabbroic affinities. More recently, Prinz et al. (1977c) have pointed out that the meteorite shows a medium to coarse-grained mosaic texture, no indication of shock metamorphism, and evidence of being a cumulate as suggested by petrofabric analysis of feldspar. A planar orientation of plagioclase tablets in Serra de Magé has also been noted by Duke (1963).

Mineralogy – Serra de Magé consists essentially of plagioclase and pyroxene. Since the rock is relatively coarse-grained and heterogeneous, proportions of the two minerals vary considerably from specimen to specimen. Thus, Duke and Silver (1967) reported a mode with 56% plagioclase and 40% pyroxene, whereas Prinz et al. (1977c) quote 64% and 34%, respectively. Normative data (Table 11) for the section investigated by the latter authors include 58.6% plagioclase and 38.7% pyroxene. The plagioclase is calcic, $Ab_{4.8}An_{95.1}Or_{0.1}$ (Table 17), and commonly twinned, with mainly polysynthetic twins with multiple twin laws. Of 142 twins investigated in 107 grains there were 39% Pericline, 28% Manebach Ala-Acline, 18% Albite, and 15% Albite-Ala twins (Prinz et al., 1977c). Pyroxenes show complex exsolution features (Figs. 2, 33), two optically homogeneous phases existing now: a "low" hypersthene, $P2_1ca$, $En_{53.8}Fs_{44.6}Wo_{1.6}$ (Table 17), as host (~ 82 vol. %) and an augite, C2/c, in four distinct enclosed habits. The several types of augite include: A1 – the most abundant,

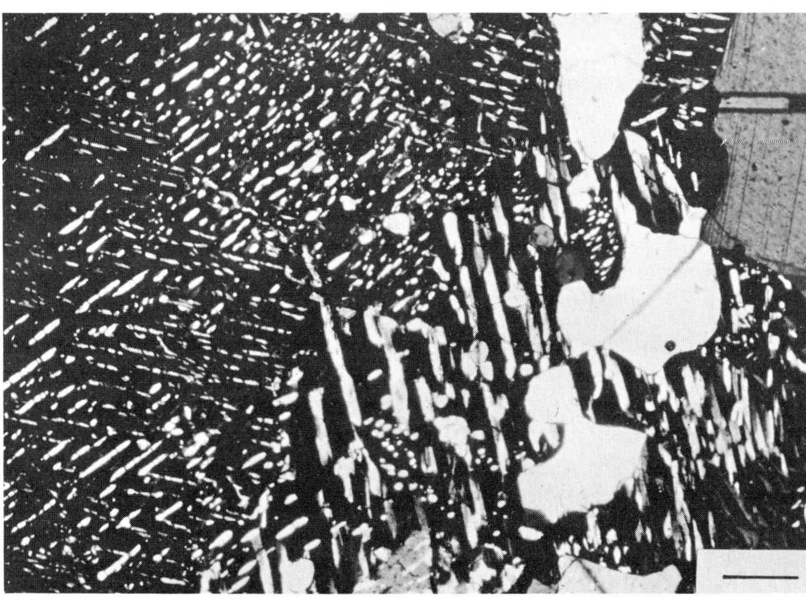

Figure 33 – Photomicrograph of the Serra de Magé, Pernambuco, Brazil monomict pigeonite-plagioclase achondrite, showing complex pyroxene exsolution features. The host is hypersthene and the lamellae are augite. Grains poikilitically enclosed in the pyroxene are plagioclase. Transmitted light, crossed polarizers. Scale bar equals 0.20 mm

thick lamellae (⩽ 25μm thick) are relict (001) exsolution from pigeonite; A2 — usually fine (< 1 μm thick) but sometimes blebby (100) lamellae from hypersthene; A3 — septa and lenses (~5 μm thick) on (052) of hypersthene; and A4 — irregular small (3-7 μm) blebs (Harlow et al., 1979). Compositional data for these types are given in Table 17. Minor constituents are chromite, $Cm_{75.7}Uv_{6.3}Sp_{7.4}Hc_{8.4}Mt_{2.2}$ (Table 17), Ni-Fe (Ni, 0.3-1.5%; Co, 0.8-1.2%), troilite, SiO_2 (tridymite; cf., Duke and Silver, 1967), ilmenite, zircon and phosphate.

Table 17. Average compositions of selected minerals in the Serra de Magé monomict pigeonite-plagioclase achondrite

	Pyroxenes*							Plagioclase	Chromite
	H	A1	A2	A3	A4	P	B		
SiO_2	52.3	52.7	52.7	53.2	53.2	52.0	52.4	44.8	0.47
TiO_2	0.24	0.37				0.23	0.26	<0.02	2.36
Al_2O_3	0.34	0.69				0.35	0.40	35.0	7.5
Cr_2O_3	0.12	0.23				0.15	0.14	n.d.	53.0
V_2O_3	n.d.	n.d.	n.d.	n.d.	n.d.	n.d.	n.d.	n.d.	0.77
Fe_2O_3									1.64+
FeO	27.4	10.5	10.6	10.8	10.5	26.3	24.4	0.12	32.6
MnO	0.95	0.42				0.88	0.85	n.d.	0.62
MgO	18.6	13.4	13.5	13.5	13.9	17.9	17.6	<0.01	1.39
CaO	0.75	22.3	22.2	22.0	22.3	2.03	4.62	19.5	0.36
Na_2O	n.d.	n.d.	n.d.	n.d.	n.d.	n.d.	n.d.	0.54	n.d.
K_2O	n.d.	n.d.	n.d.	n.d.	n.d.	n.d.	n.d.	0.02	n.d.
TOTAL	100.70	100.61	99.0	99.5	99.9	99.84	100.77	99.98	100.71
	En 53.8	37.9	38.2	38.1	38.7	52.4	50.9	Ab 4.8	Cm 75.7
	Fs 44.6	16.7	16.8	17.2	16.5	43.3	39.5	An95.1	Uv 6.3
	Wo 1.6	45.4	45.0	44.7	44.8	4.3	9.6	Or 0.1	Sp 7.4
									Hc 8.4
									Mt 2.2

+ Calculated on the basis of stoichiometry.

* H-hypersthene host; A1 to A4-augite lamellae; P-last pigeonite, mean analysis in zone of fine A2 lamella representing the average composition of combined host + augite; B-bulk or initial pigeonite, composition estimated using 18% augite and 82% hypersthene.

References:

Pyroxenes (Harlow et al., 1979; plagioclase and chromite (Prinz et al., in preparation).

n.d. = not determined.

Chemical composition — Bulk chemical data on Serra de Magé are quite variable, the largest discrepancies being the Al_2O_3, FeO, MgO, and CaO contents (Table 10). This is clearly the result of poor sampling of this relatively coarse-grained rock. The new bulk chemical analysis by E. Jarosewich in Prinz et al. (1977c) was made of 0.6 grams of the rock and resolves the wide variation in Al_2O_3 (~13-27%), yielding a more realistic value of 20.89%

Al_2O_3. Rare-earths and barium abundances (Table 18) were determined by Schnetzler and Philpotts (1969), the relatively high Eu and Ba concentrations suggesting that the meteorite is a plagioclase-rich cumulate (Fig. 29), which is borne out by the bulk major element analysis and the mineralogical composition. Additional trace element abundances are given in Table 18. Figs. 30 and 31 (Morgan et al., 1978) list lithophile, siderophile and

Table 18. Trace element abundances in the Serra de Magé monomict pigeonite-plagioclase achondrite

	(1)	(2)	(3)	(4)	(5)
Sc, ppm		14			19.2
Co, ppm		10	6.5		4.8
Ni, ppm					1
Cu, ppm			2.32		
Zn, ppm			0.78		0.64
Ga, ppm			1.42		
Ge, ppb					3.2
Se, ppb			78		105
Br, ppb			45		15
Rb, ppb					50
Ag, ppb			3.3		1.0
In, ppb			0.73		
Cd, ppb			5.3		10
Sb, ppb					0.7
Te, ppb					1.0
Cs, ppb			2.1		4.5
Ba, ppm	7.59				
Ce, ppm	1.22				
Nd, ppm	1.06				
Sm, ppm	0.327				0.3
Eu, ppm	0.298				0.5
Gd, ppm	0.47				
Dy, ppm	0.554				
Er, ppm	0.33				
Yb, ppm	0.367				
Lu, ppm					0.06
Re, ppt					0.8
Ir, ppt					2.2
Au, ppb			1.75		0.14
Tl, ppb			0.12		0.03
Bi, ppb			2.4		0.18
Th, ppb				53	
U, ppb				16	8.2

References:
(1) Schnetzler and Philpotts, 1969; (2) Schmitt et al., 1972; (3) Laul et al., 1972; (4) Morgan and Lovering, 1973; (5) Morgan et al., 1976.

chalcophile element distributions for four eucrites (Ibitira, Juvinas, Moore County and Serra de Magé). Generally, Serra de Magé (and also Moore County) tend to be lower in trace elements than Ibitira and Juvinas. Major, minor and trace element abundances in plagioclase separates and in the bulk rock have recently been measured by Ma and Schmitt (1979). Rare earth element abundances in plagioclase are La 1.2x (chondrite), Sm 0.43x, Eu 10.4x, Yb 0.16x and Lu 0.12x.

Values of 4.7 ± 0.1 and 4.0 ± 0.3 for δ ^{18}O in Serra de Magé plagioclase and pyroxene, respectively, have been determined by Taylor et al. (1965).

Nuclear particle tracks — The meteorite has a very low cosmogenic track density, ranging from 170 cm^{-2} to 3500 cm^{-2}, as measured in five samples of four different stones (Carver and Anders, 1970).

Age — A Sm-Nd internal isochron gives an age of 4.41 ± 0.02 b.y. (Lugmair et al., 1977). Rb-Sr measurements carried out by Birck and Allègre (1978) indicate an initial $^{87}Sr/^{86}Sr$ ratio of 0.69903 ± 7.

On the basis of ^{3}He, ^{21}Ne and ^{38}Ar measurements, Heymann et al. (1968) estimated a cosmic ray exposure age of 1.9 m.y., later revised to 2.5 m.y. as a result of additional determinations by Ganapathy (1970). Its ^{26}Al age, 0.065 ± 0.018 m.y., is much shorter than those determined by Heymann et al. (1968), which has been interpreted by Fuse and Anders (1969) and Carver and Anders (1970) as apparently indicative that the meteorite has suffered a two-stage irradiation.

Serra de Magé has a fission-track age of 540 ± 90 m.y. (Carver and Anders, 1970; Carver, 1974; Carver and Anders, 1976b). The meteorite has a much higher K-Ar gas-retention age of 2.7 b.y. (Heymann et al., 1968).

Origin — Serra de Magé is clearly of cumulate origin on the basis of texture (Hess and Henderson, 1949; Duke and Silver, 1967; Duke, 1969; Lovering, 1975; Prinz et al., 1977c) and incompatible-element distribution, which shows a marked positive Eu anomaly (Fig. 29; Schnetzler and Philpotts, 1969). The high U-Rb ratio and anorthite content indicate that its parent body was either relatively high in refractories or low in moderately volatile lithophiles (Morgan et al., 1978). Low pressure melting experiments on eucritic meteorites by Stolper (1977) demonstrate that Serra de Magé (and also the cumulates Moama and Moore County) may have accumulated from liquids produced by extensive fractionation of advanced partial melts of the source regions of eucritic liquids rather than from liquids similar in composition to known eucrites or their derivatives. On the other hand, Consolmagno and Drake (1977), using the experimental data collected by Stolper (1977), REE abundances, and a few assumptions, concluded that the meteorite has an origin involving accumulation from liquids produced by extensive fractionation of pyroxene and plagioclase from a Juvinas-like liquid. A detailed investigation of Serra de Magé pyroxenes (Harlow et al., 1979) shows that they initially crystallized as pigeonite but during slow cooling, exsolved augite and inverted to hypersthene. A cooling rate of 4×10^{-4} °C/y. and a crustal depth of about 5 km were estimated by those authors.

Available material — 1) Muséum National d'Histoire Naturelle, Paris, France (4 sp, 332 g); 2) Istituto di Mineralogia, Modena, Italy (3 sp, 100 g); 3) British Museum (Natural History), London, England (3 sp, 35 g); 4) American Museum of Natural History, New York, U.S.A. (2 sp, 23.8 g); 5) United States National Museum (Smithsonian Institution), Washington, D.C. (2 sp, 23.1 g); 6) Institute of Meteoritics, Albuquerque, U.S.A. (3 sp, 17.4 g); 7) Center for Meteorite Studies, Tempe, U.S.A. (1 sp, 2.0 g).

7.2. CHONDRITES

7.2.1. H-group chondrites

Avanhandava (21°27'37"S; 49°57'03"W)

Fall circumstances — The Avanhandava, State of São Paulo, meteorite fell during the evening of an undetermined day in 1952 and the fall was observed by few witnesses. The recovered stone (9.33 kg) showed a very pronounced polyhedrical shape with a greatest dimension of about 40 cm. The impact excavated a pit 102 cm deep and with a uniform diameter of 51 cm. The meteorite was listed in Hey (1966) and Hutchison et al. (1977) and a preliminary description was given by Arid and Rüegg (1965, 1969). A detailed study was carried out by Paar et al. (1976).

Texture — The meteorite exhibits a well-developed chondritic texture (Fig. 34), with chondrules about 0.1-2.0 mm in apparent size that have clearly delineated boundaries with the matrix. Internal textures of the chondrules are quite variable (e.g., radiating pyroxene, porphyritic olivine and pyroxene, barred olivine, glassy) and include turbid glass as a common phase. The matrix is fine-grained, transparent and microcrystalline.

Mineralogy — Detailed microscopic and electron microprobe study (Paar et al., 1976) indicates that olivine, $Fa_{17.3}$ (Table 19; Mason, 1967c, on the basis of optical and X-ray determinations, listed a value of Fa_{18}), low-Ca pyroxene (bronzite), $Fs_{16.7}$ (Table 20), and metallic nickel-iron (kamacite, taenite, and plessite) are the major constituents. Diopside, $En_{49.0}Fs_{5.7}Wo_{45.3}$ (Table 21), plagioclase, $Ab_{84.1}An_{10.4}Or_{5.4}$ (Table 22), chromite, $Cm_{80.2}Uv_{4.1}$

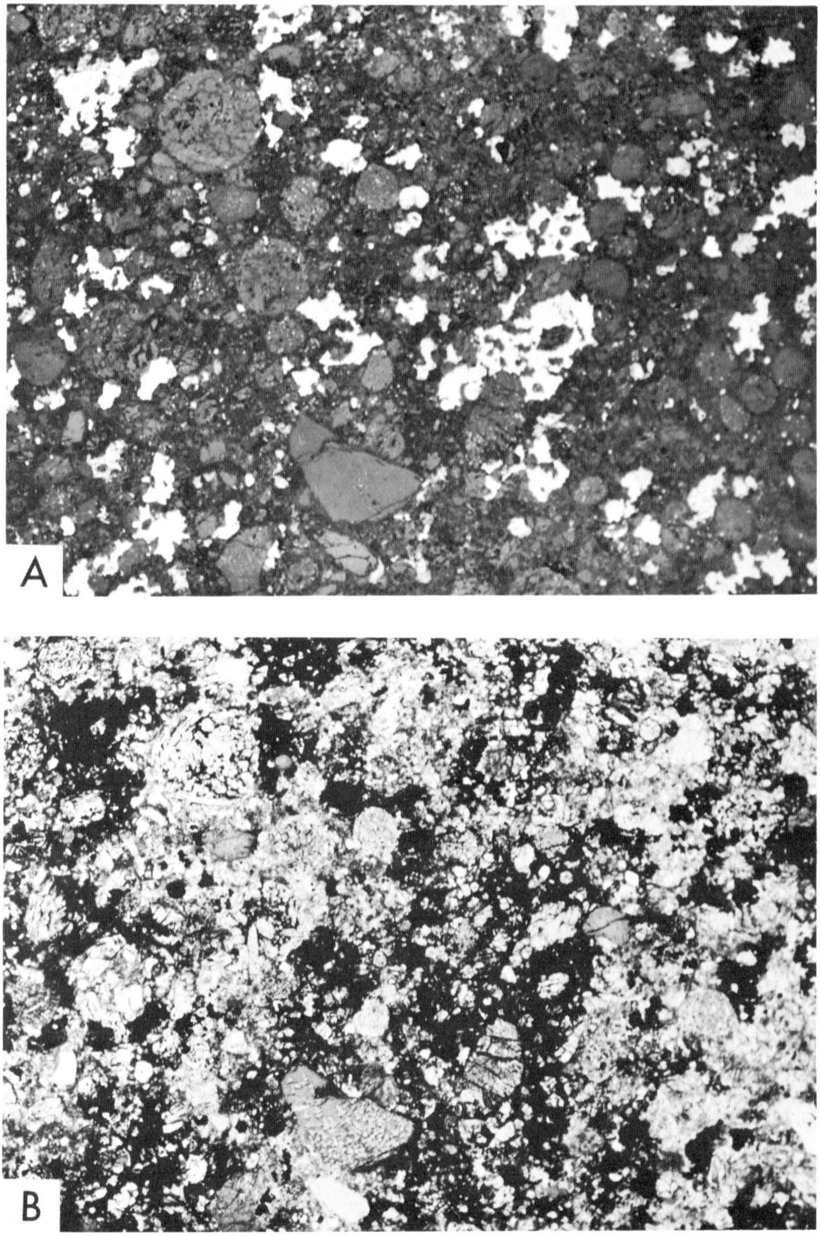

Figure 34 – Microscopic overview of the texture of the Avanhandava, São Paulo, Brazil H4 chondrite. A) Reflected light view, showing mostly silicates (olivine, pyroxene) (gray), metallic nickel-iron and troilite (white), and abundant chondrules. B) Transmitted, plane polarized light view of the same area as in Fig. 34A, showing the chondritic texture. Field of view is 7.6 mm wide.

$Pc_{0.6}Sp_{15.1}$ (Table 23), and troilite occur in minor amounts. Accessory phases include whitlockite, chlorapatite (Table 24) and the opaque minerals pentlandite, ilmenite, native copper, mackinawite, bravoite and chalcopyrrhotite.

Table 19. Average compositions of olivine from Brazilian H- and L-group chondrites

		SiO_2	TiO_2	Al_2O_3	Cr_2O_3	FeO	MnO	MgO	CaO	Total	Fo	Fa	References*
Avanhandava	H4	40.1	0.04	0.05	0.02	16.2	0.42	43.5	0.06	100.39	82.7	17.3	Paar et al. (1976)
Conquista	H4	39.6	0.02	0.07	0.03	16.2	0.46	43.7	0.02	100.10	82.8	17.2	Keil et al. (1978a)
Ipiranga	H5	38.5	n.d.	<0.01	n.d.	17.9	0.48	42.7	n.d.	99.58	81.0	19.0	Gomes et al. (1978c)
Itapicuru Mirim	H5	39.2	n.d.	0.07	n.d.	17.7	0.46	43.0	<0.02	100.43	81.2	18.8	Gomes et al. (1977b)
Macau	H5	39.4	n.d.	0.10	n.d.	18.0	0.47	42.9	<0.02	100.87	81.0	19.0	Gomes et al. (1977c)
Marília	H4	39.5	n.d.	n.d.	n.d.	17.7	n.d.	42.9	0.03	100.13	81.2	18.8	This work
São José do Rio Preto	H4	39.2	0.04	<0.01	0.02	17.6	0.50	42.8	0.02	100.18	81.3	18.7	Gomes et al. (1978d)
Sete Lagoas	H4	39.4	n.d.	0.08	n.d.	18.2	0.47	42.5	<0.02	100.65	80.6	19.4	Gomes and Keil (1977a)
Uberaba	H5	39.1	n.d.	0.09	n.d.	18.3	0.48	42.5	<0.01	100.47	80.5	19.5	Gomes et al. (1977a)
Minas Gerais	L6	38.2	n.d.	n.d.	n.d.	23.0	n.d.	38.8	0.04	100.04	75.0	25.0	This work
Paranaíba	L6	38.4	n.d.	0.08	n.d.	21.8	0.45	39.3	0.11	100.14	76.3	23.7	Keil et al. (1977)
Patrimônio	L6	38.5	n.d.	0.07	n.d.	22.7	0.45	38.7	<0.02	100.42	75.2	24.8	Gomes et al. (1978a)
Putinga	L6	38.4	n.d.	0.05	n.d.	22.7	0.44	38.7	0.02	100.31	75.2	24.8	Keil et al. (1978b)
Santa Bárbara	L4	38.3	0.02	n.d.	0.07	23.1	0.60	39.7	0.05	101.84	75.4	24.6	Berkley et al. (1978b)

* References for individual meteorites in Tables 20-23 are the same as in this table.
n.d. = not determined.

Table 20. Average compositions of low-Ca pyroxene from Brazilian H- and L-group chondrites

		SiO_2	TiO_2	Al_2O_3	Cr_2O_3	FeO	MnO	MgO	CaO	Na_2O	Total	En	Fs	Wo
Avanhandava	H4	56.1	0.12	0.13	0.09	11.2	0.46	30.9	0.68	0.04	99.72	82.0	16.7	1.3
Conquista*	H4	56.9	0.06	0.21	0.17	10.3	0.46	31.3	0.45	0.03	99.88	83.7	15.4	0.9
Ipiranga	H5	54.7	0.20	0.23	0.17	11.5	0.60	31.4	0.73	<0.01	99.53	81.8	16.8	1.4
Itapicuru Mirim	H5	56.3	n.d.	0.28	n.d.	11.4	0.47	31.0	0.79	0.08	100.32	81.7	16.8	1.5
Macau	H5	56.6	n.d.	0.25	n.d.	11.8	0.48	31.0	0.73	0.06	100.92	81.2	17.4	1.4
Marília	H4	56.5	n.d.	n.d.	n.d.	11.3	n.d.	30.8	0.65	n.d.	99.25	81.9	16.9	1.2
São José do Rio Preto	H4	55.9	n.d.	0.39	n.d.	11.4	0.46	31.1	0.65	0.08	99.98	81.9	16.8	1.2
Sete Lagoas	H4	56.1	n.d.	0.31	n.d.	11.7	0.47	31.2	0.71	0.08	100.57	81.6	17.1	1.3
Uberaba	H5	56.6	n.d.	0.25	n.d.	11.6	0.48	31.0	0.70	0.07	100.70	81.6	17.1	1.3
Minas Gerais	L6	56.0	n.d.	n.d.	n.d.	13.9	n.d.	28.9	0.87	n.d.	99.67	77.4	20.9	1.7
Paranaíba	L6	54.1	0.18	0.27	0.11	13.9	0.46	29.2	0.90	0.06	99.18	77.6	20.7	1.7
Patrimônio	L6	55.9	n.d.	0.23	n.d.	13.9	0.47	29.1	0.87	0.05	100.52	77.6	20.7	1.7
Putinga	L6	55.8	0.25	0.21	n.d.	14.1	0.52	28.6	0.84	<0.02	100.32	77.1	21.3	1.6
Santa Bárbara	L4	55.5	0.13	0.08	0.02	13.9	0.51	29.5	0.74	n.d.	100.38	78.0	20.6	1.4

* This meteorite also contains pigeonite (SiO_2 = 55.4 , TiO_2 = 0.06, Al_2O_3 = 0.33, Cr_2O_3 = 0.91, FeO = 10.5, MnO = 0.47, MgO = 30.7, CaO = 1.60, Na_2O = 0.35, Total = 100.32%; En = 81.4, Fs = 15.6, Wo = 3.0).
For references see Table 19.
n.d. = not determined.

Table 21. Average compositions of high-Ca pyroxene from Brazilian H- and L-group chondrites

		SiO_2	TiO_2	Al_2O_3	Cr_2O_3	FeO	MnO	MgO	CaO	Na_2O	Total	En	Fs	Wo
Avanhandava	H4	54.2	0.43	0.68	0.69	3.6	0.22	17.3	22.3	0.63	100.05	49.0	5.7	45.3
Itapicuru Mirim	H5	54.4	n.d.	0.71	n.d.	4.4	0.24	17.3	21.4	0.67	99.12	49.2	7.0	43.8
São José do Rio Preto	H4	53.4	0.16	0.34	0.45	4.23	0.34	18.2	21.1	n.d.	98.22	50.9	6.6	42.4
Patrimônio	L6	54.3	n.d.	0.63	n.d.	5.2	0.21	16.6	21.3	0.71	98.95	47.6	8.5	43.9
Santa Bárbara	L4	50.8	0.27	0.26	0.78	7.0	0.28	19.5	20.2	0.46	99.55	51.3	10.4	38.3

For references see Table 19; n.d. = not determined.

Table 22. Average compositions of plagioclase from Brazilian H- and L-group chondrites

		SiO_2	TiO_2	Al_2O_3	FeO	MgO	CaO	Na_2O	K_2O	Total	Ab	An	Or
Avanhandava	H4	64.7	n.d.	20.8	0.68	0.26	2.2	9.9	0.97	99.51	84.1	10.4	5.4
Ipiranga	H5	66.3	n.d.	20.9	0.69	0.45	2.39	8.96	0.83	100.52	82.8	12.2	5.0
Itapicuru Mirim	H5	64.1	n.d.	21.7	0.85	0.44	2.8	9.7	0.96	100.55	81.7	13.0	5.3
Macau	H5	64.2	n.d.	21.4	1.09	0.39	2.66	9.7	0.97	100.41	82.1	12.5	5.4
São José do Rio Preto	H4	63.6	n.d.	20.3	0.87	n.d.	3.14	12.5	0.51	100.92	85.8	11.9	2.3
Sete Lagoas	H4	64.8	n.d.	21.7	0.72	0.42	2.61	9.8	1.04	101.09	82.2	12.1	5.7
Uberaba	H5	65.0	n.d.	21.3	0.88	0.42	2.4	10.0	0.86	100.86	83.9	11.3	4.8
Paranaíba*	L6	66.2	n.d.	21.4	0.25	n.d.	2.24	9.1	0.94	100.13	83.1	11.3	5.6
Patrimônio	L6	65.3	n.d.	21.4	0.75	0.38	2.24	9.0	1.10	100.17	82.1	11.3	6.6
Putinga*	L6	65.5	n.d.	21.0	0.53	n.d.	2.57	9.3	1.14	100.04	81.0	12.4	6.6
Santa Bárbara	L4	66.3	0.02	21.9	0.64	<0.01	1.94	9.1	0.86	100.76	84.8	10.0	5.2

* Plagioclase has been converted into isotropic glass (maskelynite) by solid state shock transformation.
For references see Table 19.
n.d. = not determined.

Table 23. Average compositions of chromite from Brazilian H- and L-group chondrites

		SiO_2	TiO_2	Al_2O_3	Cr_2O_3	V_2O_3	FeO	MnO	MgO	CaO	Total	Cm	Uv	Pc	Sp
Avanhandava	H4	n.d.	1.6	7.4	55.8	0.72	30.0	0.80	3.2	n.d.	99.52	80.2	4.1	0.6	15.1
Conquista	H4	0.20	1.35	5.54	57.9	0.71	28.3	1.34	3.02	n.d.	98.36	79.8	4.3	4.2	11.7
Ipiranga	H5	n.d.	2.43	6.2	57.4	n.d.	30.4	0.83	3.11	n.d.	100.37	77.9	6.3	3.0	12.7
Itapicuru Mirim	H5	0.09	2.33	6.3	55.5	0.71	30.6	1.10	3.2	n.d.	99.83	78.2	6.3	2.7	12.8
Macau	H5	0.10	2.15	6.4	56.3	0.72	30.3	1.05	3.5	n.d.	100.52	77.3	5.8	3.9	12.9
São José do Rio Preto	H4	0.15	2.05	5.6	57.0	0.66	28.8	1.12	3.27	n.d.	98.65	76.9	6.0	5.4	11.7
Sete Lagoas	H4	0.04	2.25	6.0	55.8	0.72	30.6	1.02	3.34	n.d.	99.77	78.2	5.9	3.8	12.2
Uberaba	H5	0.10	2.4	6.2	55.7	0.69	30.5	1.09	3.4	n.d.	100.08	77.2	6.4	3.8	12.6
Paranaíba	L6	0.06*	2.80	5.5	54.6	0.65	32.4	0.92	3.03	0.09	100.05	78.9	7.3	2.9	11.0
Patrimônio	L6	0.18	2.94	5.7	54.5	0.72	32.3	0.88	2.7	n.d.	99.92	78.8	8.2	1.3	11.6
Putinga	L6	0.09	3.02	5.4	55.9	0.80	32.6	0.37	2.6	n.d.	100.78	79.0	8.2	1.8	11.0
Santa Bárbara	L4	n.d.	2.22	5.5	56.5	0.67	31.3	1.63	2.68	n.d.	100.50	81.9	5.6	1.1	11.4

* New value.
For references see Table 19.
n.d. = not determined.

Classification — Chemical and mineralogical data support classification of Avanhandava as an H-group (olivine-bronzite) chondrite. In addition to olivine and pyroxene compositions (Tables 19, 20; Fig. 35), plagioclase composition (Table 22) compares well with that given by Van Schmus and Ribbe (1968). Bulk chemical parameters (Table 10; Paar et al., 1976) that are consistent with those of H-group chondrites include Fe/SiO_2 (0.74), Fe°/Fe (0.59) and total Fe (27.15%) (Table 4). Assignment to petrologic type 4 of Van Schmus and Wood (1967) (Table 5) is based on the uniform compositions of olivine and pyroxene, the abundance of monoclinic pyroxene, and the occurrence of well-defined chondrules containing glass or microcrystalline material.

Table 24. Average compositions of phosphate minerals (in wt. %) in the Avanhandava chondrite (after Paar et al., 1976). Number of grains analyzed is shown in parentheses; n.d. = not determined

	Whitlockite (50)	Chlorapatite (35)
SiO_2	0.42	0.09
Al_2O_3	0.02	n.d.
FeO	0.61	0.42
MnO	0.01	0.01
MgO	3.9	0.17
CaO	46.0	53.9
Na_2O	2.7	0.27
P_2O_5	45.8	41.8
F	n.d.	0.6
Cl	n.d.	4.9
TOTAL	99.46	102.16
	Minus O = F, Cl	1.36
		100.80

Age — Gas retention ages of 4.3 b.y. (U-He) and 4.45 b.y. (K-Ar) were reported by Müller and Zähringer (1969).

Available material — 1) Museu Nacional, Rio de Janeiro, Brazil (1 sp, 7,672 g; Fig. 36); 2) Instituto Geológico, São Paulo, Brazil (1 sp, 725 g); 3) Faculdade de Filosofia, Ciências e Letras de São José do Rio Preto, São José do Rio Preto, Brazil (1 sp, 493.8 g); 4) United States National Museum (Smithsonian Institution), Washington, D.C. (1 sp, 135 g); 5) Instituto de Geociências, São Paulo, Brazil (1 sp, 52 g); 6) Institute of Meteoritics, Albuquerque, U.S.A. (3 sp, 38.6 g); 7) Naturhistorisches Museum Wien, Vienna, Austria (1 sp, 12 g); 8) Istituto di Mineralogia, Modena, Italy (10 g).

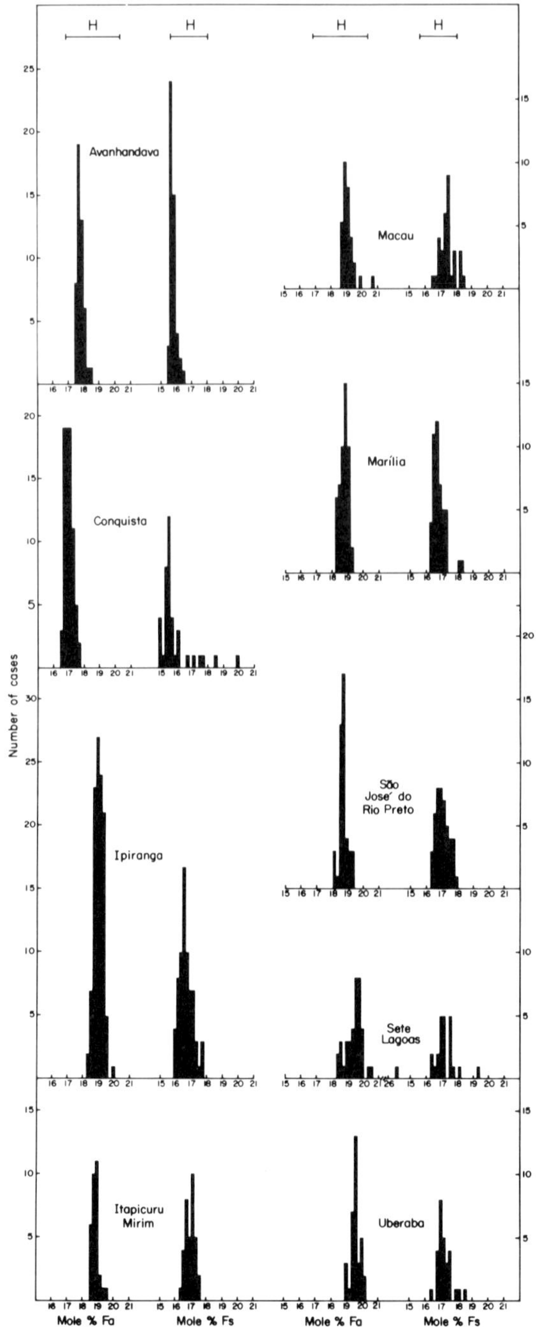

Figure 35 — Histograms showing the compositions, in terms of mole % endmembers, of olivine (Fa; Fe_2SiO_4) and low-Ca pyroxene (Fs; $FeSiO_3$) in Brazilian H-group chondrites. For comparison, ranges in average compositions of olivine and pyroxene in equilibrated H-group chondrites are given (after Keil and Fredriksson, 1964; as revised by Fodor et al., 1976). References: Avanhandava and Marília (this work); others see Table 19.

Figure 36 – Avanhandava, São Paulo, Brazil H4 chondrite, weighing 7,672 g and showing a pronounced polyhedral shape. The black fusion crust is slightly stained due to mild terrestrial weathering (Photo courtesy of Smithsonian Institution, Washington, D.C., U.S.A.).

Conquista (19°51′S; 47°33′W)

Fall circumstances – The Conquista, State of Minas Gerais, meteorite fell in the morning (6.00 hr) of an unknown day in early December, 1965, at Fazenda da Mateira, Município de Conquista. Sound effects were noticed and the meteorite was reported to have excavated a pit 1.5 m in diameter and 1 m deep. The stone apparently traveled from the S and a single specimen (approximately 20.35 kg; Figs. 37, 38) was recovered by local residents. A brief abstract on the meteorite was published by Murta (1971) and a detailed description was given by Keil et al. (1978a).

Texture – The meteorite shows a well-developed chondritic texture (Fig. 3), with chondrules, round to elongate in shape, ranging in apparent size from 0.1 to 1.9 mm (mean 0.5 mm). They also vary considerably in internal textures, and barred, excentro-radial, porphyritic, and granular varieties were observed. The matrix consists of fine-grained, microcrystalline material that resembles recrystallized glass in places. Glassy material is also found within the chondrules.

Figure 37 – Conquista, Minas Gerais, Brazil H4 chondrite, weighing approximately 20.35 kg, covered by a black fusion crust with abundant aerodynamic surface depressions.

Figure 38 – Side view of the Conquista, Minas Gerais, Brazil H4 chondrite, showing its well-developed chondritic texture.

Mineralogy — According to Keil et al. (1978a), the meteorite consists mainly of olivine, $Fa_{17.2}$ (Table 19), low-Ca pyroxene (bronzite), $Fs_{15.4}Wo_{3.0}$ (Table 20), and metallic nickel-iron, with minor amounts of troilite and glassy material of variable composition (Table 25). Accessory phases include pigeonite, $Fs_{15.6}$ (Table 20), chromite, $Cm_{79.8}Uv_{4.3}Pc_{4.2}Sp_{11.7}$ (Table 23), and high-Ca pyroxene. Hydrous ferric oxide of terrestrial origin is also present, mainly rimming the metal phase.

Table 25. Representative analyses of the alkali-SiO_2-rich glassy material from the Conquista chondrite (after Keil et al., 1978a)

	Average	Ranges
SiO_2	61.7	56.4 – 65.1
TiO_2	0.33	0.13 – 0.67
Al_2O_3	11.9	6.8 – 18.5
Cr_2O_3	0.45	0.19 – 0.96
FeO	3.9	2.6 – 6.3
MnO	0.07	0.03 – 0.21
MgO	9.0	4.0 – 17.9
CaO	7.1	1.9 – 12.8
Na_2O	5.5	3.4 – 8.2
K_2O	1.04	0.16 – 2.83
ZrO_2	0.03	<0.01 – 0.11
P_2O_5	0.20	<0.01 – 1.04
TOTAL	101.22	

Classification — Olivine and pyroxene compositions (Tables 19, 20; Fig. 35) are close to the low iron ends of the ranges for H-group chondrites. Chromite composition (Table 23) also suggests H-group classification (Table 4). H-group classification is also evident from the bulk composition (Table 10): when assuming that all Fe_2O_3 formed by terrestrial weathering and oxidation of metallic nickel-iron, the recalculated analysis shows metallic nickel-iron of 18.5%, total iron of 25.83%, and the ratios of Fe°/Ni of 9.59, Fe°/Fe of 0.64, and Fe/SiO_2 of 0.69. Assignment to the petrologic type 4 of Van Schmus and Wood (1967) (Table 5) is indicated by the slight variabilities in compositions of olivine and pyroxene, the well-defined chondritic texture, the abundance of twinned low-Ca clinopyroxene, and the presence of glassy material, particularly in the chondrules.

Available material — 1) Museu de Histório Natural, Universidade Federal de Minas Gerais, Belo Horizonte, Brazil (1 sp, 20.35 kg; Figs. 37, 38); 2) Instituto de Geociências, São Paulo, Brazil (2 sp, 87.1 g); 3) Institute of Meteoritics, Albuquerque, U.S.A. (5 sp, 50.7 g).

Ipiranga (25° 30'S; 54° 30'W)

Fall circumstances — The Ipiranga, State of Paraná, meteorite fell in the morning (10.30 hr) of December 27, 1972, at Lajeado Ipiranga, Foz do Iguaçu county. Many witnesses saw a fireball passing through a cloudless sky, leaving behind a bright trail. Sound effects (two loud detonations followed by what sounded like machine gun fire) and a smoke trail that lasted more than 15 minutes were also noticed. At Lajeado do Ipiranga and nearby villages, houses were shaken, cups tinkled, doors trembled, and windows and bookstands vibrated. Some of the stones excavated impact pits, apparently the largest one having been 45 cm in diameter and 25 cm deep. More than 30 stones of a total weight of over 7 kg were recovered by local residents. The meteorite was first described by Cavarretta et al. (1975) and is listed in Hutchison et al. (1977). More detailed descriptions of its texture and mineralogy are given by Angelucci and Funiciello (1975) and Gomes et al. (1978c).

Texture — The meteorite displays a chondritic texture (Fig. 39), with chondrules ranging in apparent diameter from 0.2 to 2.0 mm. They are round to elongate in shape and vary considerably in internal textures, and barred, excentro-radial, porphyritic and granular varieties were observed. The meteorite also shows clear textural evidences of shock effects (shock veins, wavy extinction, mechanical deformation of cleavage and twin sets, etc.).

Mineralogy — According to Angelucci and Funiciello (1975) and Gomes et al. (1978c), the meteorite consists of major olivine, $Fa_{19.0}$ (Table 19), bronzite, $Fs_{16.8}$ (Table 20) and kamacite-taenite (plessite); minor troilite and plagioclase, $Ab_{82.8}An_{12.2}Or_{5.0}$ (Table 22), and accessory chromite, $Cm_{77.9}Uv_{6.3}Pc_{3.0}Sp_{12.7}$ (Table 23). Angelucci and Funiciello (1975) also reported pentlandite and oldhamite as accessory phases. However, the occurrence of oldhamite is doubtful: this mineral has not previously been found in the relatively oxidized ordinary chondrites but only in the highly reduced enstatite chondrites and enstatite achondrites (e.g., Keil, 1968; Graham et al., 1977b). Until electron microprobe data are presented of this phase in Ipiranga, its occurrence must be considered uncertain.

Classification — Chemical and mineralogical data support classification of Ipiranga as an H-group chondrite. In addition to olivine and pyroxene compositions (Tables 19, 20; Fig. 35), plagioclase (Table 22) and chromite (Table 23) compositions compare well with those for H-group chondrites (Table 4). The bulk chemical composition (Table 10) is also consistent with H-group classification, as indicated by the ratios Fe/SiO_2 (0.71;

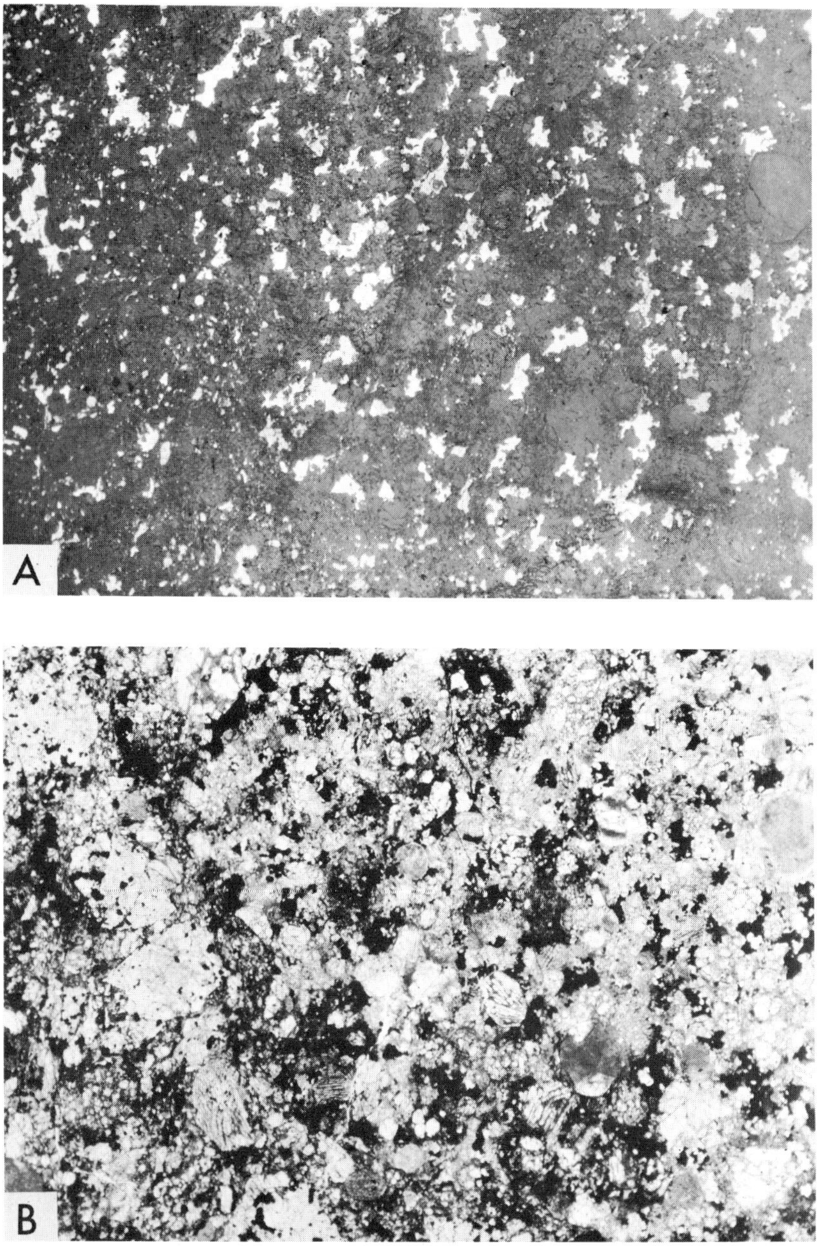

Figure 39 — Microscopic overview of the texture of the Ipiranga, Paraná, Brazil H5 chondrite. A) Reflected light view, showing mostly ferromagnesian minerals (olivine, pyroxene) (gray), metallic nickel-iron and troilite (white), and a few chondrules. B) Transmitted light view of the same area as shown in Fig. 39A, displaying a somewhat more chondritic (right) and a somewhat more recrystallized (left) portion of the meteorite. Recrystallization to the left is, at least in part, related to shock effects. Field of view is 7.6 mm wide.

0.73), Fe°/Fe (0.57; 0.54) and Fe°/Ni (8.91; 8.29) and the contents of total Fe (26.27%; 26.78%) and metallic nickel-iron (16.73%; 16.35%) (data listed first after Gomes et al., 1978c; those listed second after Angelucci and Funiciello, 1975). Textural and mineralogical evidence (i.e., the uniform and homogeneous compositions of olivine and orthopyroxene, the poorly-developed chondritic texture, and the relatively high degree of recrystallization of the matrix) indicate that the meteorite belongs to the petrologic type 5 of Van Schmus and Wood (1967) (Cavarretta et al., 1975; Angelucci and Funiciello, 1975; Gomes et al., 1978c).

Figure 40 – One specimen of the Ipiranga, Paraná, Brazil chondrite, weighing 2,650 g; the meteorite is covered by a black fusion crust and shows regmaglypts in places.

Available material – 1) Museo di Mineralogia, Rome, Italy (4 sp, 3,294 g); 2) Observatório Astronômico do Colégio Estadual do Paraná, Curitiba, Brazil (1 sp, 2,650 g; Fig. 40); 3) Instituto de Geociências, São Paulo, Brazil (1 sp, 234.2 g); 4) Institute of Meteoritics, Albuquerque, U.S.A. (3 sp, 24 g); 5) United States National Museum (Smithsonian Institution), Washington, D.C. (3 sp, 11.1 g); 6) Center for Meteorite Studies, Tempe, U.S.A. (1 sp, 3.2 g).

Itapicuru Mirim (3°24'S; 44°20'W)

Fall circumstances — The Itapicuru Mirim, State of Maranhão, meteorite fell in the morning (11.00 hr) on a clear day in March, 1879. The fall was accompanied by sound effects (shot gun) and one single stone (2,024 g) was reported to be recovered (Fig. 41). It exhibits a black fusion crust and also shallow surface depressions. The meteorite was listed by Derby (1888a, b), Oliveira (1931), Vidal (1936) and Hey (1966), and a detailed description was given by Gomes et al. (1977b).

Texture — The meteorite has a chondritic texture (Fig. 42), with perfectly round to elongated chondrules ranging from 0.3 up to 1.5 mm (mean 0.6 mm). Many different textural types of chondrules were observed, including barred, excentro-radial, porphyritic, and granular varieties. Chondrules and chondrule fragments are embedded into a fine-grained, microcrystalline matrix.

Figure 41 — Itapicuru Mirim, Maranhão, Brazil H5 chondrite, now weighing 1,301 g, is covered by a black fusion crust with a few contraction cracks.

Mineralogy — Mineralogical and electron microprobe investigations (Gomes et al., 1977b) show olivine, $Fa_{18.8}$ (Table 19; a value of Fa_{18} was reported by Mason, 1963, on the basis of optical and X-ray determinations), low-Ca pyroxene (bronzite), $Fs_{16.8}$ (Table 20), and metallic nickel-iron (kamacite and taenite) as major minerals, with minor amounts of plagioclase, $Ab_{81.7}An_{13.0}Or_{5.3}$ (Table 22) and troilite, and accessory amounts of high-Ca pyroxene (diopside), $En_{49.2}Fs_{7.0}Wo_{43.8}$ (Table 21), chromite, $Cm_{78.2}Uv_{6.3}Pc_{2.7}Sp_{12.8}$ (Table 23), native copper, pentlandite and whitlockite.

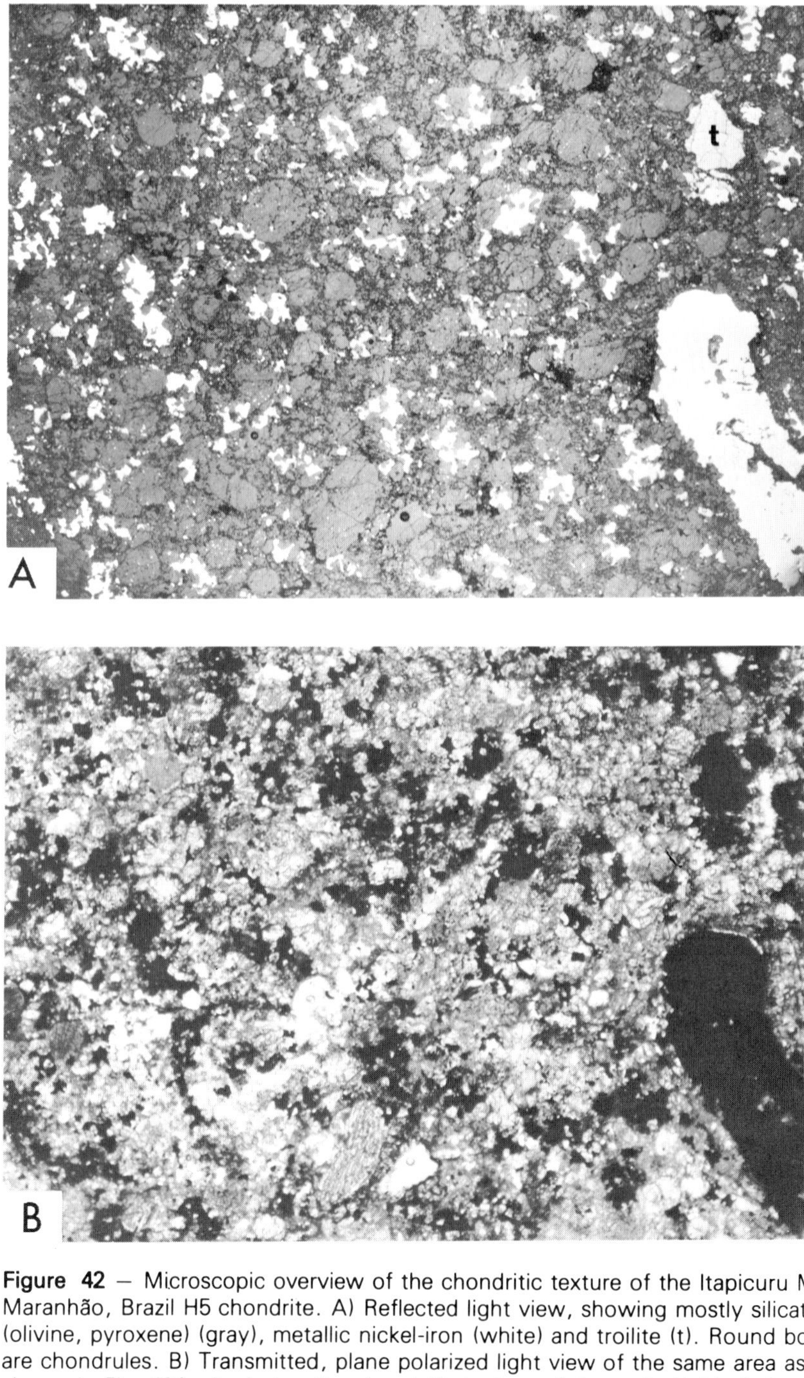

Figure 42 – Microscopic overview of the chondritic texture of the Itapicuru Mirim, Maranhão, Brazil H5 chondrite. A) Reflected light view, showing mostly silicates (olivine, pyroxene) (gray), metallic nickel-iron (white) and troilite (t). Round bodies are chondrules. B) Transmitted, plane polarized light view of the same area as shown in Fig. 42A, displaying the chondritic texture of the rock. Field of view is 7.6 mm wide.

Classification — On the basis of mineral compositions, particularly olivine and low-Ca pyroxene (Tables 19, 20; Fig. 35), and chromite (Table 23), it is concluded that the Itapicuru Mirim meteorite belongs to the H-group (Table 4). Bulk chemical data, particularly the ratios Fe°/Ni (9.89), Fe/SiO_2 (0.83) and Fe°/Fe (0.65) as well as the contents of total Fe (29.58%) and metallic nickel-iron (21.07%) (Table 10) are also consistent with an H-group classification (Table 4). On the basis of the uniformity and homogeneity of ferromagnesian minerals, notably olivine and pyroxene, the absence of igneous glass, and the degree of recrystallization of the matrix, the meteorite is placed into petrologic class 5 of Van Schmus and Wood (1967) (Table 5).

Available material — 1) Museu Nacional, Rio de Janeiro, Brazil (1 sp, 1,301 g; Fig. 41); 2) Field Museum of Natural History, Chicago, U.S.A. (1 sp, 220 g); 3) Instituto de Geociências, São Paulo, Brazil (3 sp, 110.8 g); 4) Istituto di Mineralogia, Modena, Italy (20 g); 5) United States National Museum (Smithsonian Institution), Washington, D.C. (2 sp, 16.6 g); 6) Humboldt University, Berlin, German Democratic Republic (1 sp, 7 g); 7) British Museum (Natural History), London, England (1 sp, 6.3 g); 8) Muséum National d'Histoire Naturelle, Paris, France (1 sp, 6 g); 9) Yale University, New Haven, U.S.A. (1 sp, 5.4 g); 10) Institute of Meteoritics, Albuquerque, U.S.A. (2 sp, 3.7 g).

Macau (5°12'S; 36°40'W)

Synonyms — Macao, Macayo.

Fall circumstances — The Macau, State of Rio Grande do Norte, meteorite fell in the afternoon (5.00 hr) of November 11, 1836. The meteorite fall, a shower consisting of individual stones of the size of doves' eggs, was accompanied by sound (detonations) and light phenomena (a brilliant fireball visible up to 600 km away from the place of fall). Several cattle were reported to have been killed by the fall. The largest concentration of stones was reported near the mouth of the river Assu, but material was scattered over an extensive area. However, apparently only four stones (2,047 g) were recovered by local residents. The meteorite was listed by Derby (1888a, b), Oliveira (1931), Vidal (1936), Hey (1966) and Hutchison et al. (1977), and a detailed description was given by Gomes et al. (1977c).

Texture — The texture of Macau is clearly chondritic, although recrystallization has caused chondrule-matrix boundaries to be diffuse (Fig. 43). The chondrules are rounded and sometimes elongated in shape and range in apparent size from 0.2 up to

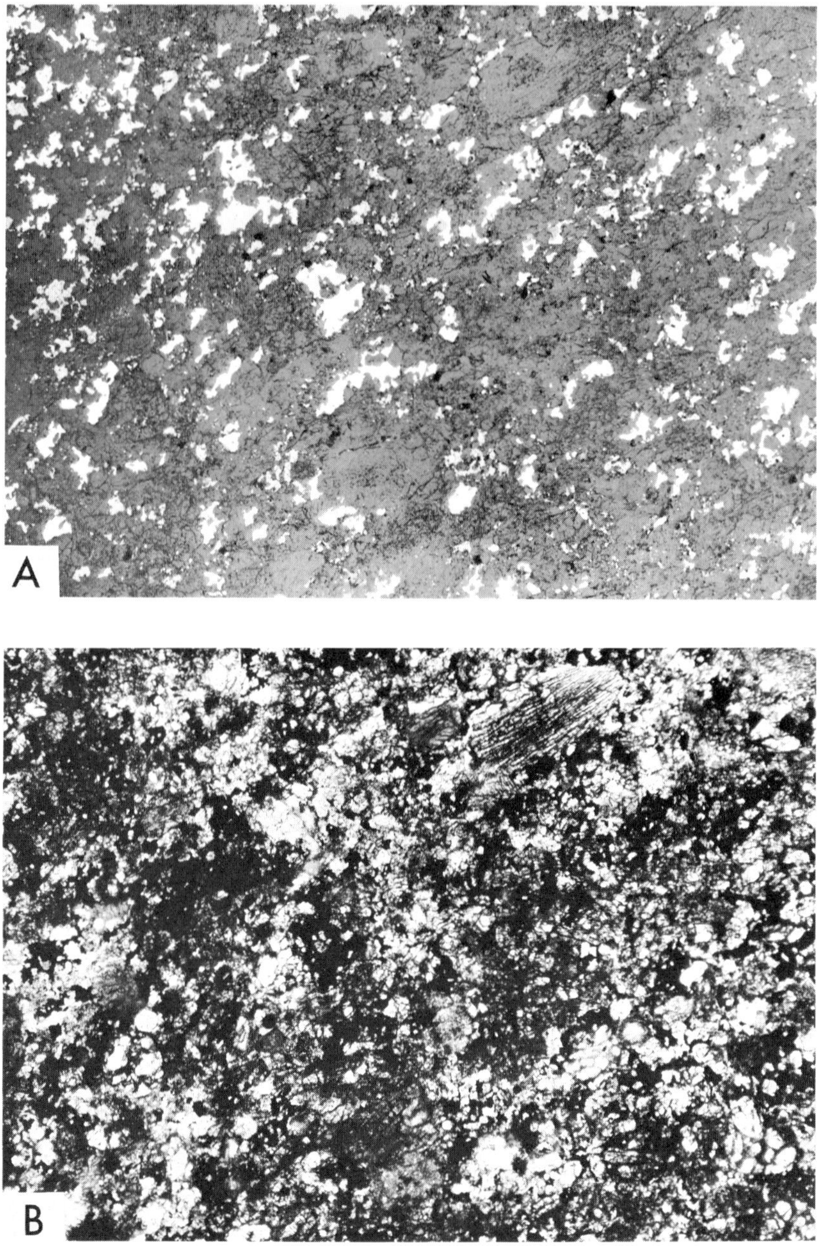

Figure 43 — Microscopic overview of the poorly-defined chondritic texture of the Macau, Rio Grande do Norte, Brazil H5 chondrite. A) Reflected light view, showing mostly ferromagnesian silicates (olivine, pyroxene) (gray), metallic nickel-iron (white) and troilite (light-gray). B) Transmitted, plane polarized light view of the same area as shown in Fig. 43A, displaying a few relict chondrules. Field of view is 7.6 mm wide.

2.4 mm (mean 0.6 mm). Their internal textures are quite variable, including barred, porphyritic and excentro-radial varieties. The chondrules are either monosomatic or polysomatic.

Mineralogy — Mineralogical and electron microprobe studies by Gomes et al. (1977c) show olivine, $Fa_{19.0}$ (Table 19; an identical value was listed by Mason, 1963, on the basis of optical and X-ray determinations), bronzite, $Fs_{17.4}$ (Table 20), and metallic nickel-iron (kamacite and taenite) as major constituents, with minor amounts of plagioclase, $Ab_{82.1}An_{12.5}Or_{5.4}$ (Table 22), and troilite. Accessory minerals include chromite, $Cm_{77.3}Uv_{5.8}Pc_{3.9}Sp_{12.9}$ (Table 23), high-Ca pyroxene (diopside), and hydrous ferric oxide of terrestrial origin, mainly associated with the metal phase.

Classification — On the basis of its mineral compositions, particularly of olivine and pyroxene (Tables 19, 20; Fig. 35) and chromite (Table 23), it is concluded that Macau belongs to the H-group chondrites (Table 4). Although the meteorite is somewhat oxidized by terrestrial weathering, the bulk chemical data (Table 10) also support an H-group classification. If one assumes that all Fe_2O_3 formed by weathering of metallic nickel-iron, the recalculated analysis shows typical H-group characteristics: total Fe is 26.27%,

Figure 44 — A specimen of the Macau, Rio Grande do Norte, Brazil H5 chondrite weighing 359.8 g. It is almost entirely covered with fusion crust and shows some evidence for terrestrial weathering.

total metallic nickel-iron is 18.14%, and the ratios are Fe/SiO_2 = = 0.72, Fe°/Ni = 9.76, and Fe°/Fe = 0.62 (Table 4). The uniform and homogeneous compositions of olivine and pyroxene, the only poorly-developed chondritic texture, and the degree of recrystallization of the matrix indicate that the meteorite belongs to the petrologic type 5 of Van Schmus and Wood (1967) (Table 5), in agreement with the earlier results of these authors.

Available material — 1) Naturhistorisches Museum Wien, Vienna, Austria (6 sp, 591 g); 2) Museu Nacional, Rio de Janeiro, Brazil (1 sp, 359.8 g; Fig. 44); 3) Muséum National d'Histoire Naturelle, Paris, France (4 sp, 284 g); 4) Field Museum of Natural History, Chicago, U.S.A. (3 sp, 84 g); 5) United States National Museum (Smithsonian Institution), Washington, D.C. (3 sp, 76.1 g); 6) Center for Meteorite Studies, Tempe, U.S.A. (1 sp, 42.2 g); 7) Humboldt University, Berlin, German Democratic Republic (37 g; cf., Hey, 1966); 8) American Museum of Natural History, New York, U.S.A. (1 sp, 22.5 g); 9) Meteorite Committee of the USSR, Moscow (1 sp, 15.8 g); 10) Istituto di Mineralogia, Modena, Italy (15 g); 11) Harvard University, Cambridge, U.S.A. (1 sp, 15 g); 12) Instituto de Geociências, São Paulo, Brazil (1 sp, 14.2 g); 13) Institute of Meteoritics, Albuquerque, U.S.A. (3 sp, 13.0 g); 14) Max-Planck-Institut für Chemie, Mainz, Federal Republic of Germany (2 sp, 10.4 g); 15) Göttingen University, Göttingen, Federal Republic of Germany (1 sp, 10.0 g); 16) Yale University, New Haven, U.S.A. (1 sp, 9.2 g); 17) British Museum (Natural History), London, England (2 sp, 6 g).

Marília (22°15'S; 49°56'W)

Fall circumstances — The Marília, State of São Paulo, meteorite shower fell in the afternoon (5.00 hr) of October 5, 1971, and was observed by many witnesses. At least seven individual specimens were collected, the largest one weighing 780 g. The total weight of the recovered material is estimated to be 2.5 kg. In spite of the fall having taken place within the city limits of Marília, no additional information on the circumstances of the fall are available. The first description of the meteorite was given by Avanzo et al. (1973); the meteorite is also listed in Hutchison et al. (1977).

Texture — The meteorite shows a pronounced chondritic texture (Fig. 45), with chondrules of variable sizes and shapes. They exhibit well-defined contours and are embedded in a fine-grained matrix containing microcrystalline material. The chondrules vary considerably in internal texture, and porphyritic, barred, fibrous, and granular types were observed.

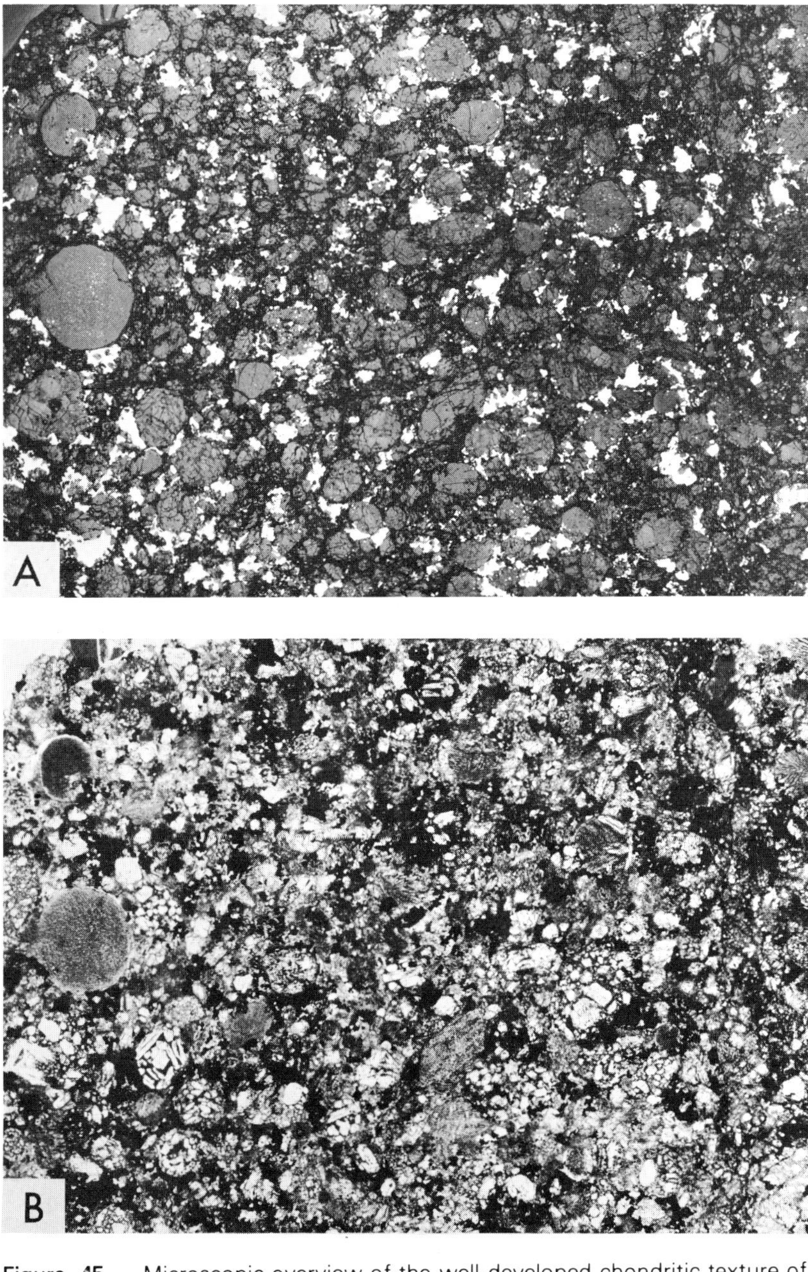

Figure 45 — Microscopic overview of the well-developed chondritic texture of the Marília, São Paulo, Brazil H4 chondrite. A) Reflected light view, showing mostly ferromagnesian minerals (olivine, pyroxene) (gray), metallic nickel-iron and troilite (white), as well as well-defined chondrules. B) Transmitted, plane polarized light view of the same area as shown in Fig. 45A, illustrating the chondritic texture with chondrules of several textural types. Field of view is 7.6 mm wide.

Mineralogy – The meteorite consists mainly of olivine, pyroxene, nickel-iron and troilite, with minor amounts of plagioclase and chromite (Avanzo et al., 1973). Probe data on olivine and low-Ca pyroxene (this work) average $Fa_{18.8}$ (Table 19) and $Fs_{16.9}$ (Table 20), respectively.

Classification – On the basis of mineral compositions, particularly of olivine and low-Ca pyroxene (Tables 19, 20; Fig. 35), the meteorite belongs to the H-group (Table 4). This classification is confirmed by bulk chemical analyses (Table 10), particularly the ratios Fe/SiO_2 (0.75, 0.68), Fe°/Fe (0.61, 0.62) and Fe°/Ni (9.87, 9.72) and the contents of total Fe (27.75%, 26.16%) and metallic nickel-iron (18.59%, 17.80%) (data listed first from previously unpublished analysis by Jarosewich; those listed second from Shima et al., 1974). Additional support for H-group classification comes from the abundances of heavy trace elements (Table 26) which, according to Shima et al. (1974), are consistent with the values given in the literature for H-group chondrites. Table 26 also lists some trace element abundances in the metallic nickel-iron phases of the meteorite. Assignment to petrologic type 4 of Van Schmus

Table 26. Trace element abundances in the Marília chondrite

	Bulk meteorite* (1)	Metallic nickel-iron † (2)
Ni		7.6
Co		4390
As		10.8
Ru		5.1
W	265	0.94
Re	125	0.43
Os	2030	4.4
Ir	1240	3.7
Pt	2810	9.0
Au	440	1.07
Hg	740	
Tl	5.2	
Pb	150	
Bi	2.8	
Th	56.5	
U	17.5	

* Values are in ppb.
† Except for Ni (%), values are in ppm.
References:
(1) Shima et al., 1974; (2) Rambaldi, 1977; for W, Rambaldi and Cendales, 1977.

and Wood (1967) (Table 5) is suggested by the slight variabilities in compositions of olivine and pyroxene, the well-developed chondritic texture, and the presence of abundant twinned, low-Ca clinopyroxene.

Age — Activities of radionuclides (^{54}Mn, ^{22}Na, ^{60}Co, ^{26}Al and ^{40}K), as measured by gamma-ray spectrometry on two specimens (106 g and 780 g), were reported by Nordemann and Sakai (1973). Interpretation of the measurements gave the following results: mass of the meteorite in space, $\geqslant$ 50 kg; cosmic ray exposure age, $\geqslant 1.3 \times 10^6$ years.

Available material — 1) Instituto de Geociências, Salvador, Brazil (1 sp, 780 g); 2) Instituto de Geociências, São Paulo, Brazil (10 sp, 610 g); 3) Istituto di Mineralogia, Modena, Italy (200 g); 4) Institute of Meteoritics, Albuquerque, U.S.A. (1 sp, 57.7 g); 5) Center for Meteorite Studies, Tempe, U.S.A. (1 sp, 11.6 g); 6) Naturhistorisches Museum Wien, Vienna, Austria (1 sp, 7 g); 7) United States National Museum (Smithsonian Institution), Washington, D.C. (1 sp, 5.7 g).

São José do Rio Preto (20°48′36″S; 49°22′50″W)

Fall circumstances — The São José do Rio Preto, State of São Paulo, meteorite fell in the morning (8.00 hr) of August 14, 1962 (Fazenda Urtiga), about 12 km W of São José do Rio Preto, and was observed by at least one witness. Apparently the meteorite traveled from the SW to the NE and one single stone (927 g) was reported to have been recovered. The meteorite is listed in Hey (1966) and Hutchison et al. (1977) and a preliminary description was given by Coutinho and Arid (1963). It was described in detail by Gomes et al. (1978d).

Texture — The meteorite shows a well-developed chondritic texture (Fig. 46). Chondrules are abundant and vary considerably in internal textures (barred, excentro-radial, and porphyritic varieties). They are clearly discernible from the groundmass, a fine-grained, microcrystalline material, which appears to be recrystallized glass in places.

Mineralogy — Detailed microscopic and electron microprobe study (Gomes et al., 1978d) shows olivine, $Fa_{18.7}$ (Table 19; on the basis of optical and X-ray diffraction techniques, Mason, 1967c, reported a value of Fa_{18}), bronzite, $Fs_{16.8}$ (Table 20), and metallic nickel-iron (kamacite and taenite) as major minerals, with minor amounts of troilite, plagioclase, $Ab_{85.8}An_{11.9}Or_{2.3}$ (Table 22), high-Ca pyroxene (diopside), $En_{50.9}Fs_{6.6}Wo_{42.4}$ (Table 21), and chromite, $Cm_{76.9}Uv_{6.0}Pc_{5.4}Sp_{11.7}$ (Table 23).

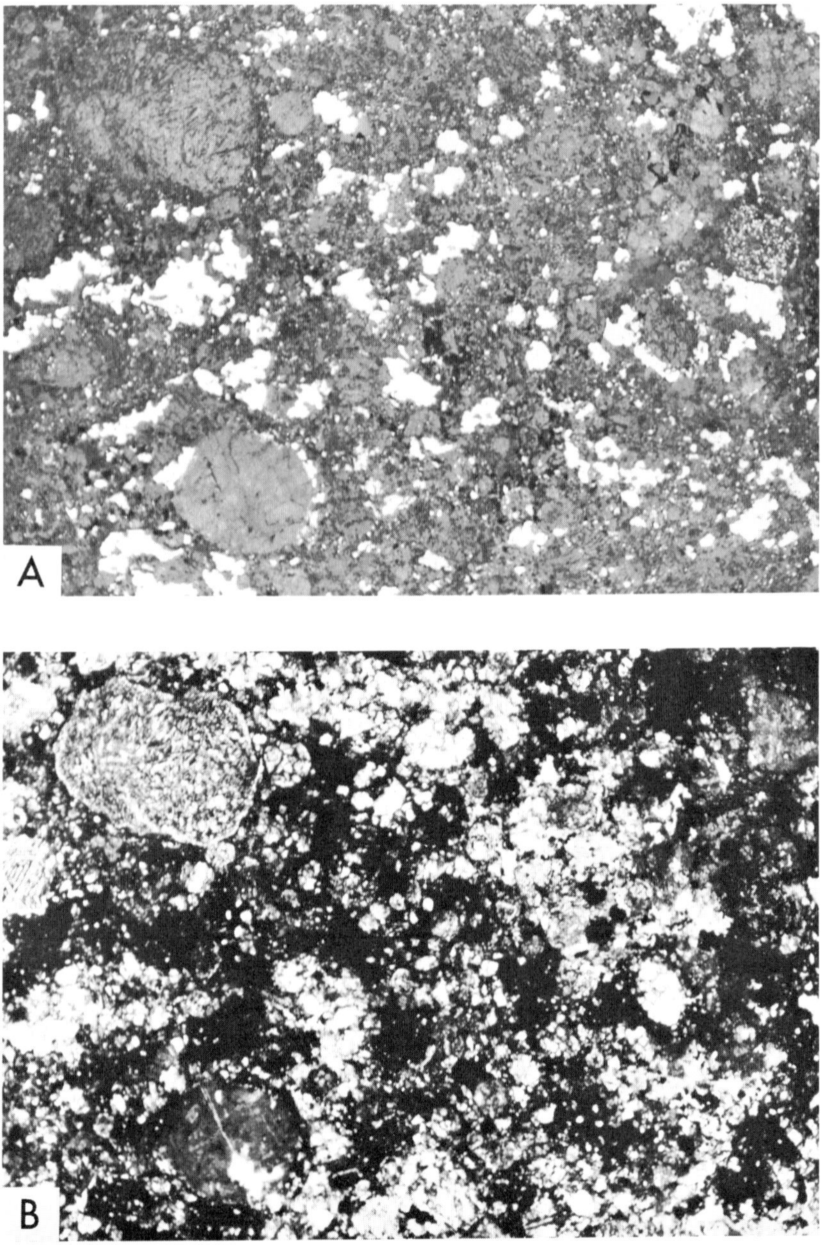

Figure 46 – Microscopic overview of the texture of the São José do Rio Preto, São Paulo, Brazil H4 chondrite. A) Reflected light view, showing mostly ferromagnesian silicates (gray), metallic nickel-iron (white), as well as well-developed chondrules. B) Transmitted, plane polarized light view of the same area as shown in Fig. 46A, illustrating the pronounced overall chondritic texture, with chondrules of various textural types. Field of view is 7.6 mm wide.

Classification — On the basis of the composition of its constituent minerals, particularly olivine, pyroxene (Tables 19, 20; Fig. 35) and chromite (Table 23), the meteorite clearly belongs to the H-group (Table 4). This classification is also supported by the bulk chemical data (Table 10), particularly the ratios of Fe°/Ni (9.40), Fe/SiO_2 (0.73) and Fe°/Fe (0.59) as well as the contents of total Fe (26.78%) and total metallic nickel-iron (17.64%) (Table 4). The moderate variability in the composition of olivine and pyroxene and the well-developed chondritic texture, the presence of abundant fine-grained microcrystalline material in chondrules and in the matrix, and the common occurrence of twinned low-Ca pyroxene indicates that the meteorite belongs to petrologic type 4 of Van Schmus and Wood (1967) (Table 5).

Figure 47 — São José do Rio Preto, São Paulo, Brazil H4 chondrite, weighing 564.8 g, covered by black fusion crust with numerous contraction cracks.

Available material — 1) Faculdade de Filosofia, Ciências e Letras de São José do Rio Preto, São José do Rio Preto, Brazil (1 sp, 564.8 g; Fig. 47); 2) Instituto de Geociências, São Paulo, Brazil (1 sp, 80.1 g); 3) Institute of Meteoritics, Albuquerque, U.S.A. (4 sp, 27.7 g); 4) United States National Museum (Smithsonian Institution), Washington, D.C. (2 sp, 16.0 g); 5) Istituto di Mineralogia, Modena, Italy (10 g).

Sete Lagoas (19°28'S; 44°13'W)

Fall circumstances – The Sete Lagoas, State of Minas Gerais, meteorite fell on December 15, 1908, and the fall was observed by only a few witnesses. Sound effects were noticed, and four small and slightly rounded stones, the largest about the size of a walnut, were recovered by local people (63 g total weight according to Hey, 1966). However, there are presently 14 stones in the collection of the Escola de Minas, Ouro Preto, most of which are covered totally by melting crust (Fig. 48). Thus, more than four individual stones must have been associated with this fall. Furthermore, we account for more than 135 g of the meteorite in the world's meteorite collections. The meteorite is also listed in Oliveira (1931) and Hutchison et al. (1977), and a detailed description was given by Gomes and Keil (1977a).

Figure 48 – Fourteen stones of the Sete Lagoas, Minas Gerais, Brazil H4 chondrite weighing a total of 75.3 g. The largest specimen is 3.8 cm in longest dimension.

Texture – The meteorite exhibits a well-developed chondritic texture (Fig. 49). Chondrules are quite variable in internal textures (barred, excentro-radial, and porphyritic types) and range in apparent size from 0.1 to 2.0 mm (mean 0.6 mm). Chondrules and chondrule fragments are embedded in a fine-grained interstitial material that is glassy in places.

Mineralogy – According to Gomes and Keil (1977a), the meteorite consists mainly of olivine, $Fa_{19.4}$ (Table 19; Mason,

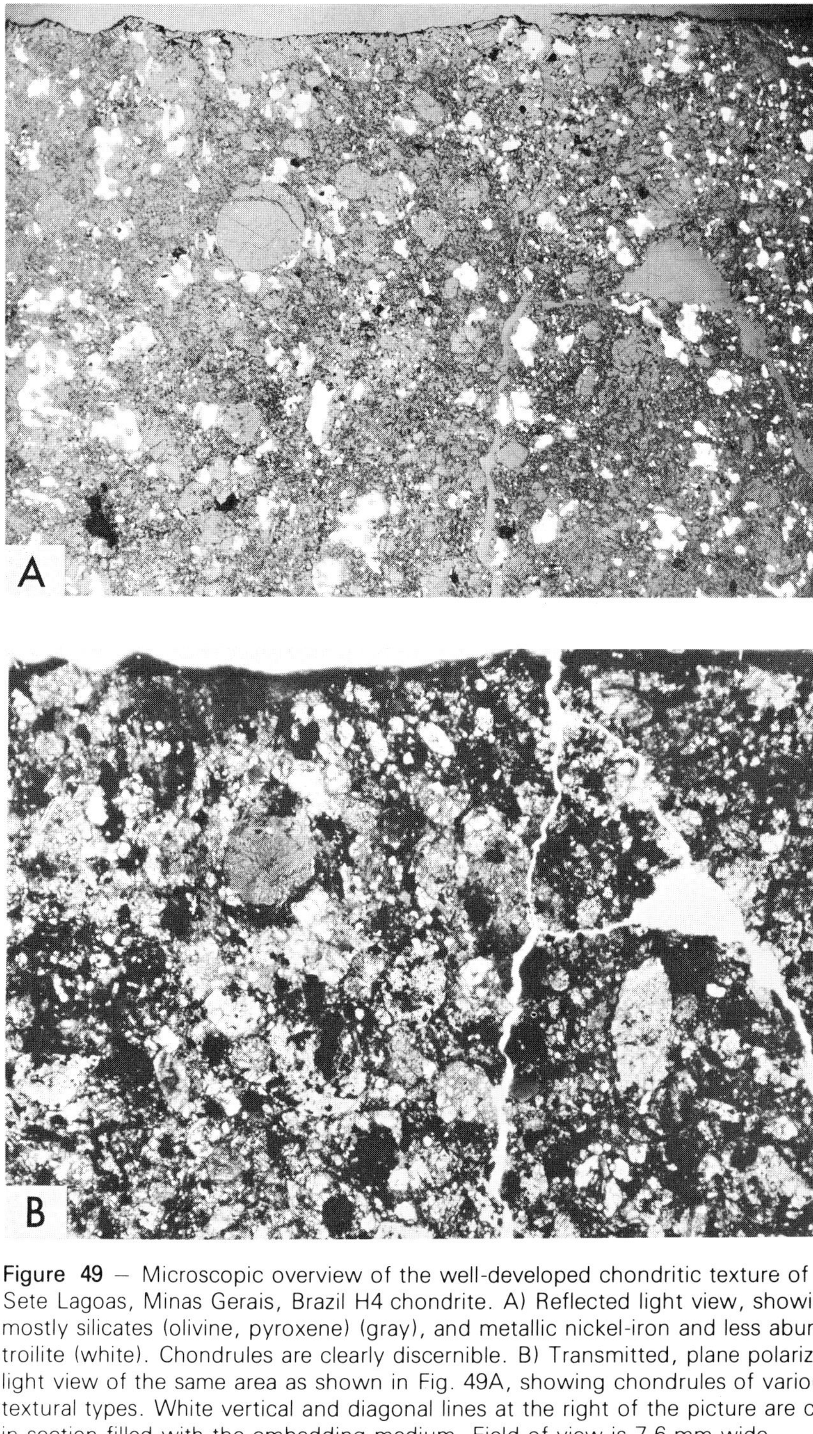

Figure 49 — Microscopic overview of the well-developed chondritic texture of the Sete Lagoas, Minas Gerais, Brazil H4 chondrite. A) Reflected light view, showing mostly silicates (olivine, pyroxene) (gray), and metallic nickel-iron and less abundant troilite (white). Chondrules are clearly discernible. B) Transmitted, plane polarized light view of the same area as shown in Fig. 49A, showing chondrules of various textural types. White vertical and diagonal lines at the right of the picture are cracks in section filled with the embedding medium. Field of view is 7.6 mm wide.

1963, listed a value of Fa_{19} as determined by optical and X-ray diffraction methods), low-Ca pyroxene (bronzite), $Fs_{17.1}$ (Table 20), and metallic nickel-iron (kamacite and taenite), with subordinate amounts of plagioclase, $Ab_{82.2}An_{12.1}Or_{5.7}$ (Table 22), and troilite. Accessory minerals are chromite, $Cm_{78.2}Uv_{5.9}Pc_{3.8}Sp_{12.2}$ (Table 23), whitlockite, and a gray (reflected light) material, principally forming thin rims around metallic nickel-iron grains. This material is interpreted as a hydrous ferric oxide formed by terrestrial weathering.

Classification — Mineral data, particularly olivine and low-Ca pyroxene compositions (Tables 19, 20; Fig. 35), indicate an H-group classification for the meteorite. Plagioclase and chromite compositions (Tables 22, 23) also conform to those given for H-group chondrites (Table 4). The well-developed chondritic texture, the slight compositional variability of olivine and pyroxene, and the presence of some interstitial glassy material (particularly in the chondrules), place the meteorite into the petrologic type 4 of Van Schmus and Wood (1967) (Table 5).

Available material — 1) Escola de Minas, Ouro Preto, Brazil (14 sp, 75.3 g; Fig. 48); 2) Istituto di Mineralogia, Modena, Italy (40 g); 3) Institute of Meteoritics, Albuquerque, U.S.A. (4 sp, 16.9 g); 4) American Museum of Natural History, New york, U.S.A. (1 sp, 3.5 g).

Uberaba (19° 49′S; 48° 47′W)

Synonyms — Dores dos Campos Formosos, Dores do Campo.

Fall circumstances — The Uberaba, State of Minas Gerais, meteorite fell in the morning (10.00 hr) of June 29, 1903 (Fazenda do Capão Grosso, Dores dos Campos Formosos district), about 84 km from Uberaba. There is some confusion in the literature regarding the day of the fall (June 29, cf., Hussak, 1904; June 22, cf., Oliveira, 1931; June 12, cf., Vidal, 1936), but probably the date given by Hussak (1904) is the correct one, since he was the first to give a preliminary description of the meteorite, including a detailed account of the field investigations. According to witnesses, sound (detonations) and light (luminous meteor traveling from the NE to the SW) effects were noticed and a stone of about 30–40 kg was seen to fall. However, Hussak (1904) reported that a considerable portion of the stone was destroyed by local people. A detailed study of the meteorite was published by Gomes et al. (1977a).

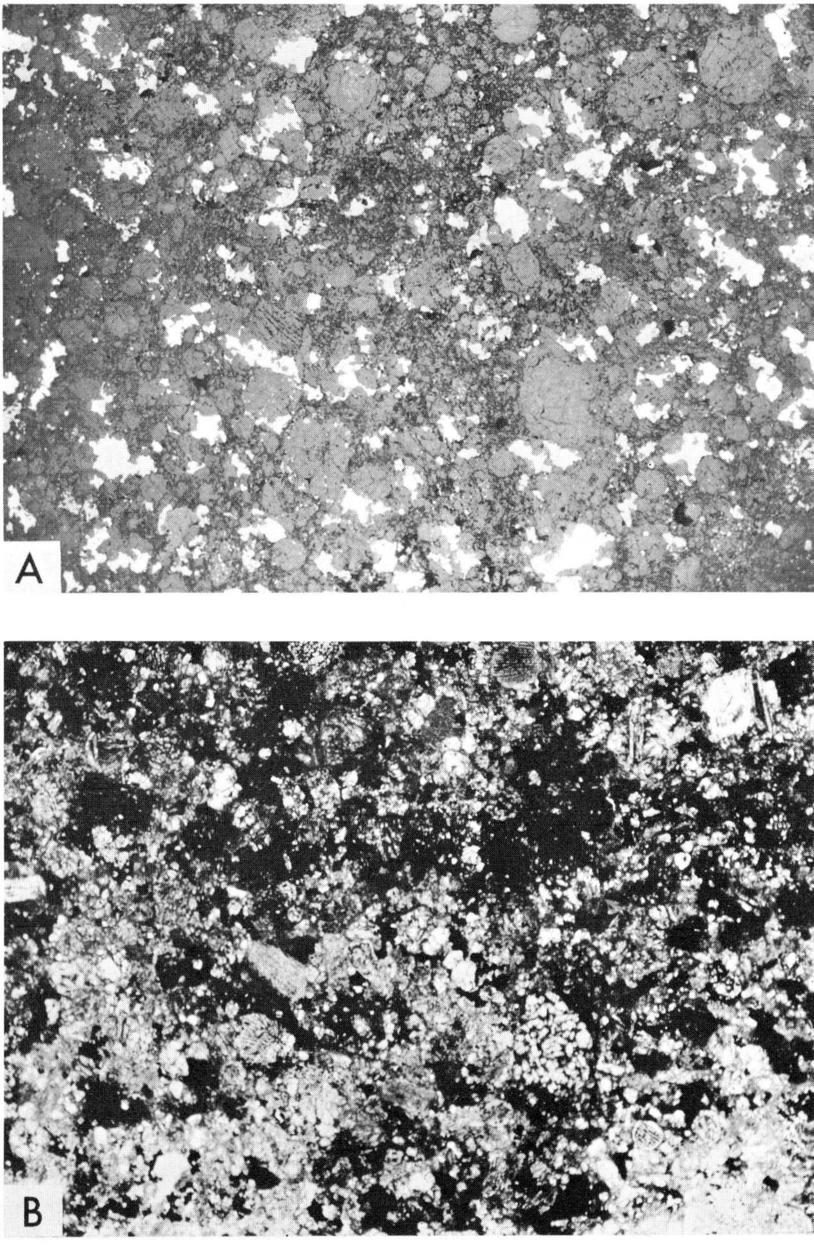

Figure 50 — Microscopic overview of the chondritic texture of the Uberaba, Minas Gerais, Brazil H5 chondrite. A) Reflected light view, showing mostly silicates (olivine, pyroxene) (gray), metallic nickel-iron and troilite (white). Round bodies are chondrules. B) Transmitted, plane polarized light view of the same area as in Fig. 50A, showing chondrules of different textural types. Field of view is 7.6 mm wide.

Texture — The meteorite exhibits a chondritic texture (Fig. 50), with chondrules, perfectly round to elongate in shape, ranging in apparent size from 0.2–1.1 mm (mean 0.5 mm). Their internal textures are variable (barred, excentro-radial, porphyritic, and granular types), with polysomatic chondrules dominating over monosomatic ones. The matrix is fine-grained, mainly consisting of microcrystalline material.

Mineralogy — Optical and electron microprobe investigations by Gomes et al. (1977a) show olivine, $Fa_{19.5}$ (Table 19; on the basis of optical and X-ray determinations, Mason, 1963, reported a value of Fa_{19}), low-Ca pyroxene (bronzite), $Fs_{17.1}$ (Table 20), and metallic nickel-iron (kamacite and taenite) as major constituents of the meteorite. Minor minerals are plagioclase, $Ab_{83.9}\,An_{11.3}\,Or_{4.8}$ (Table 22), and troilite, whereas chromite, $Cm_{77.2}\,Uv_{6.4}\,Pc_{3.8}\,Sp_{12.6}$ (Table 23), and high-Ca pyroxene (diopside) make up the accessory phases. In some places, the metal phase is rimmed by a hydrous ferric oxide of terrestrial origin.

Classification — Chemical and mineral data, particularly compositions of olivine and pyroxene (Tables 19, 20; Fig. 35), support classification of Uberaba as an H-group chondrite. Also, chromite composition (Table 23) compares well with that given by Bunch et al. (1967) for H-group chondrites (Table 4). H-group

Figure 51 — Uberaba, Minas Gerais, Brazil H5 chondrite weighing 4,735 g. It is covered by black fusion crust and shows some regmaglypts.

classification is further supported by bulk chemical data (Table 10), particularly the ratios Fe°/Ni (7.80), Fe/SiO_2 (0.70) and Fe°/Fe (0.52), as well as the contents of total Fe (26.08%) and total metallic nickel-iron (15.28%) (Table 4). Slightly lower than average H-group Fe°/Ni and Fe°/Fe ratios and total-metal values are due to terrestrial weathering of metallic nickel-iron to hydrous ferric oxide. The chondritic texture and the uniformity and homogeneity of olivine and pyroxene suggest that the meteorite Uberaba belongs to the petrologic class 5 of Van Schmus and Wood (1967) (Table 5), in agreement with the earlier results by these authors.

Available material — 1) Escola de Minas, Ouro Preto, Brazil (1 sp, 4,735 g; Fig. 51); 2) Muséum National d'Histoire Naturelle, Paris, France (5 sp, 1,167 g); 3) Naturhistorisches Museum Wien, Vienna, Austria (6 sp, 430 g); 4) Field Museum of Natural History, Chicago, U.S.A. (1 sp, 139.2 g); 5) United States National Museum (Smithsonian Institution), Washington, D.C. (3 sp, 106.4 g); 6) British Museum (Natural History), London, England (2 sp, 82 g); 7) American Museum of Natural History, New York, U.S.A. (2 sp, 72.9 g); 8) Harvard University, Cambridge, U.S.A. (1 sp, 62 g); 9) Istituto di Mineralogia, Modena, Italy (45 g); 10) Institute of Meteoritics, Albuquerque, U.S.A. (2 sp, 20.4 g); 11) Instituto de Geociências, São Paulo, Brazil (1 sp, 14.5 g); 12) Center for Meteorite Studies, Tempe, U.S.A. (1 sp, 9.6 g); 13) Max-Planck-Institut für Chemie, Mainz, Federal Republic of Germany (1 sp, 6.2 g).

7.2.2. L-group chondrites

Mafra (26° 10'S; 49° 56'W)

Fall circumstances — The Mafra, State of Santa Catarina, meteorite fell in 1941, and the fall was observed by many eyewitnesses. Sound and light phenomena were noticed before its fragmentation. Four pieces were recovered by local residents, the largest one weighing about 600 g. The samples were reported to have been buried at depths of up to 1 m. The meteorite is listed in Hey (1966) and Hutchison et al. (1977), and a preliminary description of its mineralogy was given by Valarelli and Arruda (1965). More detailed studies of the rock have been carried out by Melfi (1965) and Levi-Donati et al. (1976).

Texture — Mafra exhibits a well-developed chondritic texture, with subangular to subrounded chondrules that have complex internal textures (porphyritic, barred, excentro-radial, etc.) (Fig. 52). Some chondrules appear to have been implaced while still partially molten as they conform to adjacent chondrules. The matrix of the chondrules is turbid and, in some cases, is of a

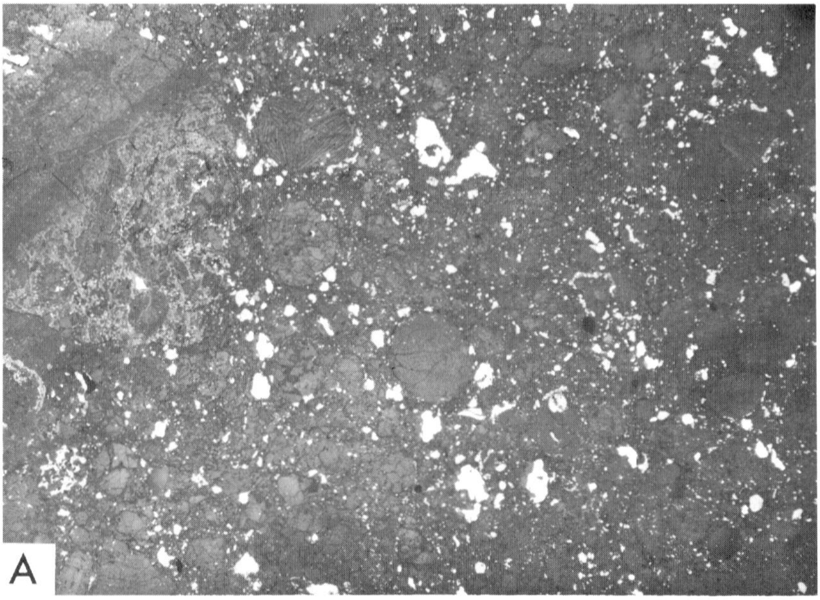

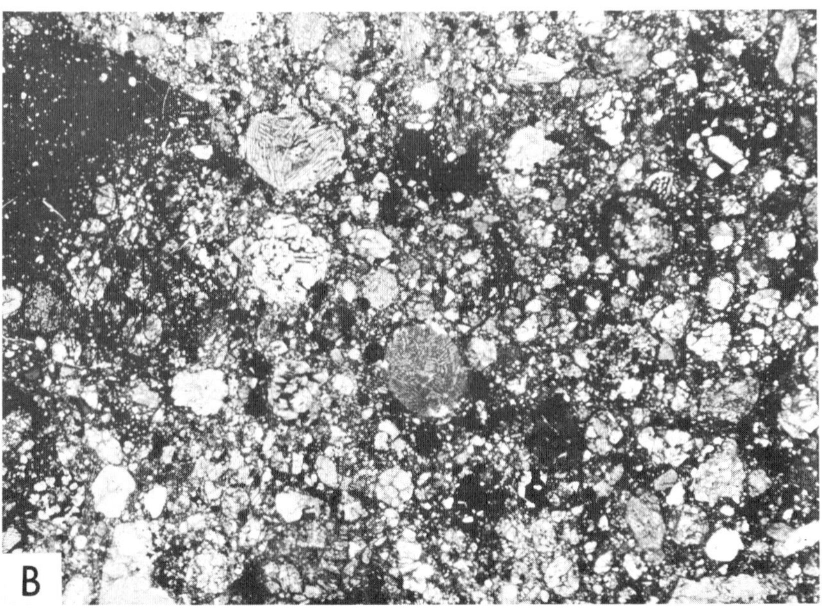

Figure 52 – Microscopic view of the Mafra, Santa Catarina, Brazil L 3-4 chondrite, showing a well-developed chondritic texture and a lithic fragment in the upper left hand corner. The clast has been moderately brecciated, and cracks are filled by a fine network of opaque minerals, notably troilite. A) Reflected light view, showing silicates (gray) and opaques (white). B) Transmitted, plane polarized light view of the same area as shown in Fig. 52A, mainly emphasizing the complex internal texture of the chondrules. Field of view is 7.6 mm wide.

burgundy color. Small crystals (1-5 μm) are commonly present in this turbid glass. The matrix of the meteorite is crystalline but fine-grained. The meteorite also contains many lithic clasts which, in general, have a similar texture to the rest of the rock. The boundaries of the clasts are usually sharp and distinct. Some clasts show clear evidence of being moderately shocked as indicated by undulatory extinction of some minerals and the presence of black shock-veins.

Mineralogy — Melfi (1965) has pointed out that the silicate and metallic fractions constitute approximately 69% and 31% (in vol.) of the whole rock, respectively. According to Levi-Donati et al. (1976), the meteorite mainly consists of olivine, pyroxene (both ortho and monoclinic phases present), and nickel-iron alloys, and minor amounts of troilite. Accessory minerals include chromite, ilmenite, and rare plagioclase. The occurrence of daubréelite has also been reported by Melfi (1965) on the basis of X-ray measurements. However, in view of the overall oxidized nature of the rock, the occurrence of this phase is questionable.

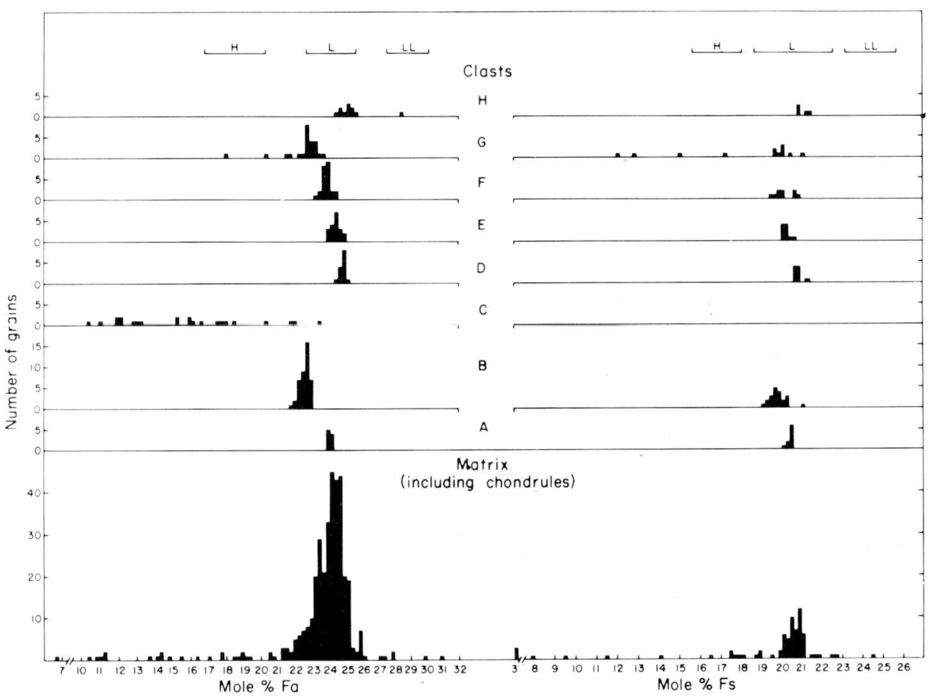

Figure 53 — Histograms illustrating the compositions of olivine and low-Ca pyroxene, as expressed as mole % endmembers fayalite (Fa; Fe_2SiO_4) and ferrosilite (Fs; $FeSiO_3$), in the Mafra, Santa Catarina, Brazil L 3-4 chondrite matrix (including chondrules) and lithic fragments A, B, C, etc. For comparison, the ranges for olivine and pyroxene compositions in H-, L-, and LL-group equilibrated chondrites are shown (after Keil and Fredriksson, 1964; as revised by Fodor et al., 1976).

Mason (1967c) listed a value of Fa_{25} for the olivine composition on the basis of optical and X-ray determinations. On the other hand, measurements by Levi-Donati et al. (1976) indicated that the mineral is not homogeneous in composition, and instead has two peaks at Fa_{20} and Fa_{25}. In spite of some compositional variation, probe data obtained for seven lithic clasts and meteorite matrix including chondrules (Lange et al., in preparation) (Fig. 53) cluster in the range of Fa_{22-25}, the average probably being in the neighborhood of Fa_{24}. Pyroxene also shows compositional variation, the data mainly ranging from Fs_{19} to Fs_{21}. An average composition around Fs_{20} is indicated (Fig. 53).

Classification — Mafra has been classified as an H-group chondrite by Levi-Donati et al. (1976) on the basis of bulk chemical data (Table 10), notably the Fe/SiO_2 (0.70), Fe°/Fe (0.55), and Fe°/Ni (8.79) ratios and the contents of total Fe (26.70%) and total metal (16.48%) (Table 4). However, L-group classification is suggested by the olivine composition (Fa_{25}, cf., Mason, 1967c; Fa_{25}, dominant peak, cf., Levi-Donati et al., 1976) (Table 4). Probe data for olivine and pyroxene (Fig. 53) are clearly consistent with L-group classification for the meteorite (Table 4). Thus, the bulk chemical analysis given by Levi-Donati et al. (1976) must be erroneous, unless it refers to a material taken from a very unusual portion of the meteorite showing H-group characteristics (e.g., a lithic fragment). Nevertheless, this possibility seems to be remote since probe data by Lange et al. (in preparation) were obtained on three sections with clasts from different parts of the meteorite. Textural (the well-defined chondrules, the presence of glass) and mineralogical (the abundance of polysynthetically twinned clinopyroxene, the scarcity of plagioclase) evidence place the meteorite into the petrologic transitional types 3 and 4 of Van Schmus and Wood (1967) (Table 5). All of the clasts belong to a higher petrologic type (all type 4, one type 5). Thus, Mafra is a genomict breccia of L-group composition, with the host material being transitional from type 3-4, and the clasts being of type 4 and type 5.

Available material — 1) Instituto de Geociências, São Paulo, Brazil (2 sp, 357.7 g); 2) Istituto di Mineralogia, Modena, Italy (100 g); 3) Institute of Meteoritics, Albuquerque, U.S.A. (1 sp, 27.3 g); 4) United States National Museum (Smithsonian Institution), Washington, D.C. (1 sp, 8.1 g).

Minas Gerais (18° 30′S; 44° 00′W)

Fall circumstances — The Minas Gerais meteorite has been known since 1888. It was found without label among specimens of the Museu Nacional collection in Rio de Janeiro and believed to have been brought from the State of Minas Gerais. The original

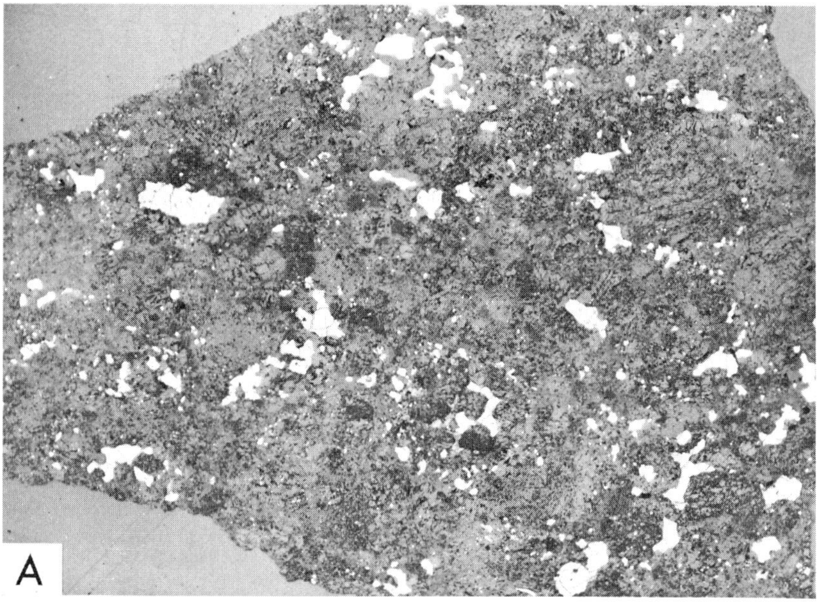

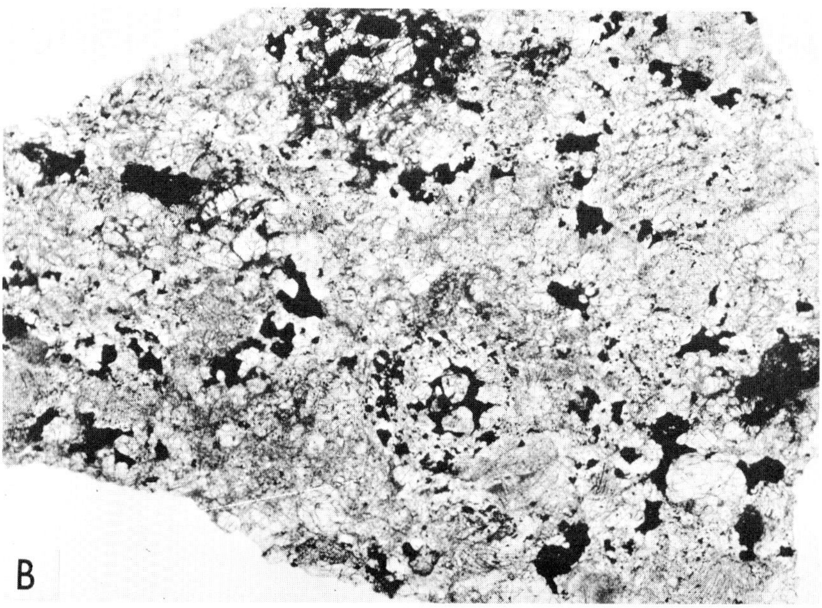

Figure 54 – Microscopic overview of the Minas Gerais, Brazil L6 chondrite. A) Reflected light view, showing mostly ferromagnesian silicates (olivine, pyroxene) (gray), metallic nickel-iron and troilite (white), and only vague outlines of chondrules. B) Transmitted, plane polarized light view of the same area as shown in Fig. 54A, showing a highly recrystallized texture with a few relict chondrules that are barely recognizable.

piece showed a light color and weighed 1,224 g (Derby, 1888a, b). According to Oliveira (1931), the Minas Gerais fall took place in 1888; he classified the meteorite as a veined, white chondrite. The stone is also listed in Hey (1966) and Hutchison et al. (1977).

Texture — The meteorite exhibits a highly-recrystallized texture (Fig. 54), with only a few chondrules still discernible from the matrix. They are round in shape and variable in internal texture (porphyritic, poikilitic, etc.). The rock also shows mineralogical and textural evidence of being slightly shocked.

Mineralogy — Keil (1962a, b) studied polished sections of Minas Gerais and reported (in weight %) 87.96 silicates, 8.03 nickel-iron, 3.75 troilite and 0.26 chromite as constituents of the rock. The normative composition (Table 11) indicates 84.9%, 8.5%, 5.3% and 0.9% of those phases, respectively. Ramdohr (1973) described chalcopyrrhotite, mackinawite, and ilmenite as accessory phases. Olivine, Fa_{25} (Table 19) and low-Ca pyroxene, $Fs_{20.9}$ (Table 20), with minor amounts of clear, interstitial plagioclase, represent the silicate portion of the meteorite. Mason (1963), on the basis of X-ray diffraction and microscopic studies, also reported olivine with Fa_{25} from the meteorite.

Classification — L-group classification for the Minas Gerais chondrite was first suggested by Keil (1962b). Our compositional data on olivine and low-Ca pyroxene (Tables 19, 20; Fig. 55) support this classification (Table 4). This classification is also confirmed by bulk chemical data (Table 10), notably the ratios Fe/SiO_2 (0.57), Fe°/Fe (0.32), and Fe°/Ni (5.28), and the contents of total Fe (22.31%) and metallic nickel-iron (8.47%) (Table 4). Textural evidence (the highly recrystallized texture of the matrix, with clear, interstitial grains of plagioclase; the poorly defined chondritic texture) and the uniform and homogeneous compositions of olivine and pyroxene, are consistent with a classification as petrologic type 6 of Van Schmus and Wood (1967) (Table 5).

Age — A $U/^4He$ gas-retention age of 3.4 b.y. for Minas Gerais was determined by Müller and Zähringer (1969).

Available material — 1) Field Museum of Natural History, Chicago, U.S.A. (1 sp, 422 g); 2) British Museum (Natural History), London, England (2 sp, 65.5 g); 3) Instituto de Geociências, São Paulo, Brazil (2 sp, 43.0 g); 4) Yale University, New Haven, U.S.A. (2 sp, 28.3 g); 5) Harvard University, Cambridge, U.S.A. (1 sp, 25.0 g); 6) United States National Museum (Smithsonian Institution), Washington, D.C. (1 sp, 10.6 g); 7) American Museum of Natural History, New York U.S.A. (2 sp, 6.2 g); 8) Naturhistorisches Museum Wien, Vienna, Austria (2 sp, 6 g); 9) Muséum National d'Histoire Naturelle, Paris, France (1 sp, 4 g).

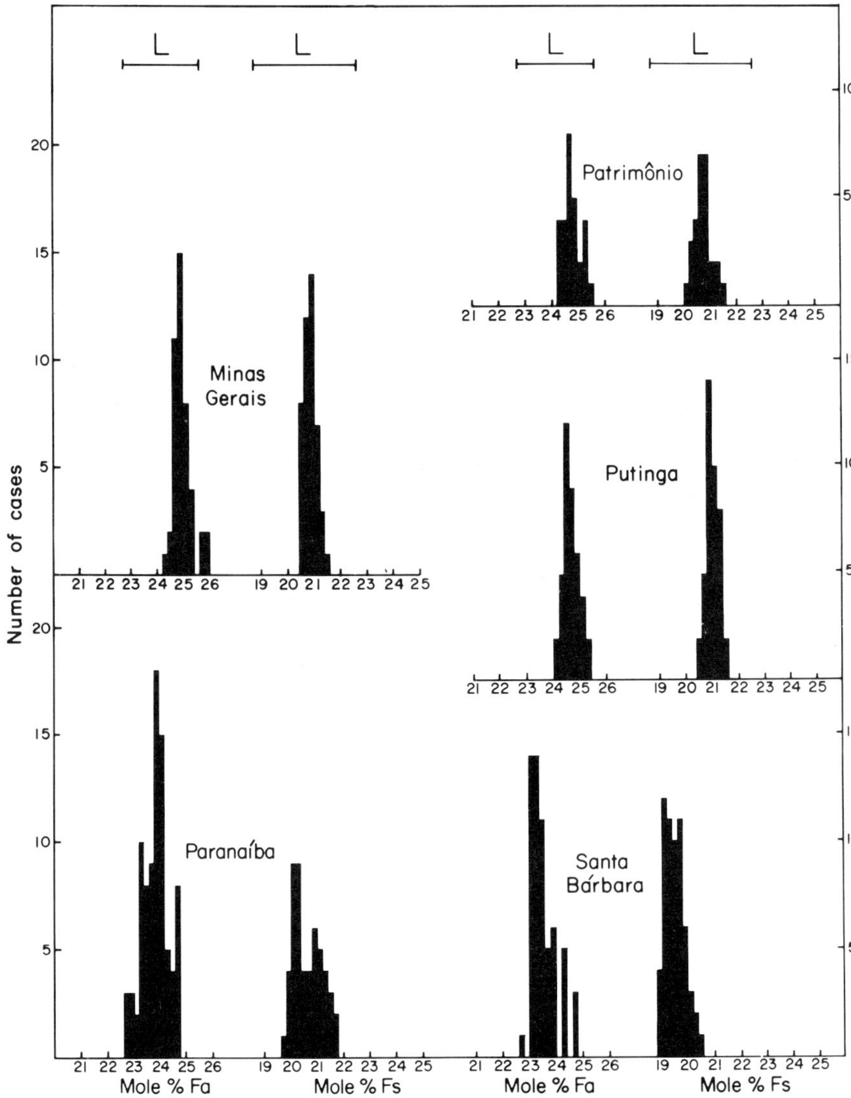

Figure 55 — Histograms showing the compositions, in terms of mole % endmembers, of olivine (Fa; Fe_2SiO_4) and low-Ca pyroxene (Fs; $FeSiO_3$) in Brazilian L-group chondrites. For comparison, ranges in average compositions of olivine and pyroxene in equilibrated L-group chondrites are given (after Keil and Fredriksson, 1964; as revised by Fodor et al., 1976). References: Minas Gerais (this work); others see Table 19.

Paranaíba (19°8'S; 51°40'W)

Synonyms — Can-Can, Mato Grosso.

Fall circumstances — The Paranaíba, State of Mato Grosso, meteorite fell in 1956 (village of Sant'Ana), about 70 km NW of Paranaíba and 80 km from the mouth of the river Aporé, and was observed by many witnesses. Sound effects were noticed and the stone was reported to have excavated a pit 2 m deep. Approximately 100 kg were recovered, but probably a considerable portion of this material was destroyed by local residents. The meteorite is listed in Hey (1966) and Hutchison et al. (1977), and a preliminary description of its texture and mineralogy was given by Arruda (1962) and Amaral (1962). Ramdohr (1973) identified chalcopyrrhotite and mackinawite as accessory phases, and a more detailed investigation of the meteorite was carried out by Keil et al. (1977) and Gomes and Keil (1978).

Texture — In hand specimen, the meteorite exhibits a very conspicuous structure, mainly consisting of light fragments, rounded to angular, quite variable in size, cut by dark veins and broad black areas (Figs. 56, 57). The light areas are highly recrystallized, with few chondrules discernible from the groundmass (Fig. 57). The dark areas were clearly derived from the light material by shock metamorphism, melted in part, and show no chondrules. The dark color of the veins results from the abundance of very fine-grained opaque phases. Thus, the meteorite is a monomict breccia of shock origin. Mineralogical considerations by Keil et al. (1977) indicate that the black shock veins formed at pressures > 430 kb and temperatures of 1500-1700° C, whereas the light fragments represent the less altered portions of the rock and experienced shock pressures of ~ 200 kb and temperatures of < 900° C.

Mineralogy — According to Keil et al. (1977), the light areas of the meteorite, which represent the less shock-altered portions of the rock, consist mainly of olivine, $Fa_{23.7}$ (Table 19; on the basis

Table 27. Isotropic grains of olivine and orthopyroxene compositions (averages) from dark shock veins of the Paranaíba chondrite (after Keil et al., 1977); n.d. = not determined

	Olivine	Pyroxene
SiO_2	38.9	54.4
TiO_2	0.02	0.18
Al_2O_3	0.11	0.26
Cr_2O_3	0.05	0.12
FeO	22.5	14.4
MnO	0.44	0.48
MgO	39.1	29.0
CaO	0.10	1.05
Na_2O	n.d.	0.08
TOTAL	101.22	99.97
	Fo 75.6	En 76.7
	Fa 24.4	Fs 21.4
		Wo 2.0

Figure 56 — Overview of a polished thin section of the Paranaíba, Mato Grosso, Brazil L6 chondrite, showing the recrystallized texture of the light areas and a shock vein (upper right hand corner), consisting of glass, isotropic and anisotropic olivine and low-Ca pyroxene, and opaque minerals. A) Reflected light view, showing silicates (gray) and opaques (white). The shock vein is clearly recognizable by the finely-disseminated, remelted metallic nickel-iron and troilite intergrowths. B) Transmitted, plane polarized light view of the same area as shown in Fig. 56A. The shock vein is dark in color due to the dissemination of tiny metallic nickel-iron and troilite grains. Note the granular olivine relict chondrule in the lower right hand corner. Field of view is 7.6 mm wide.

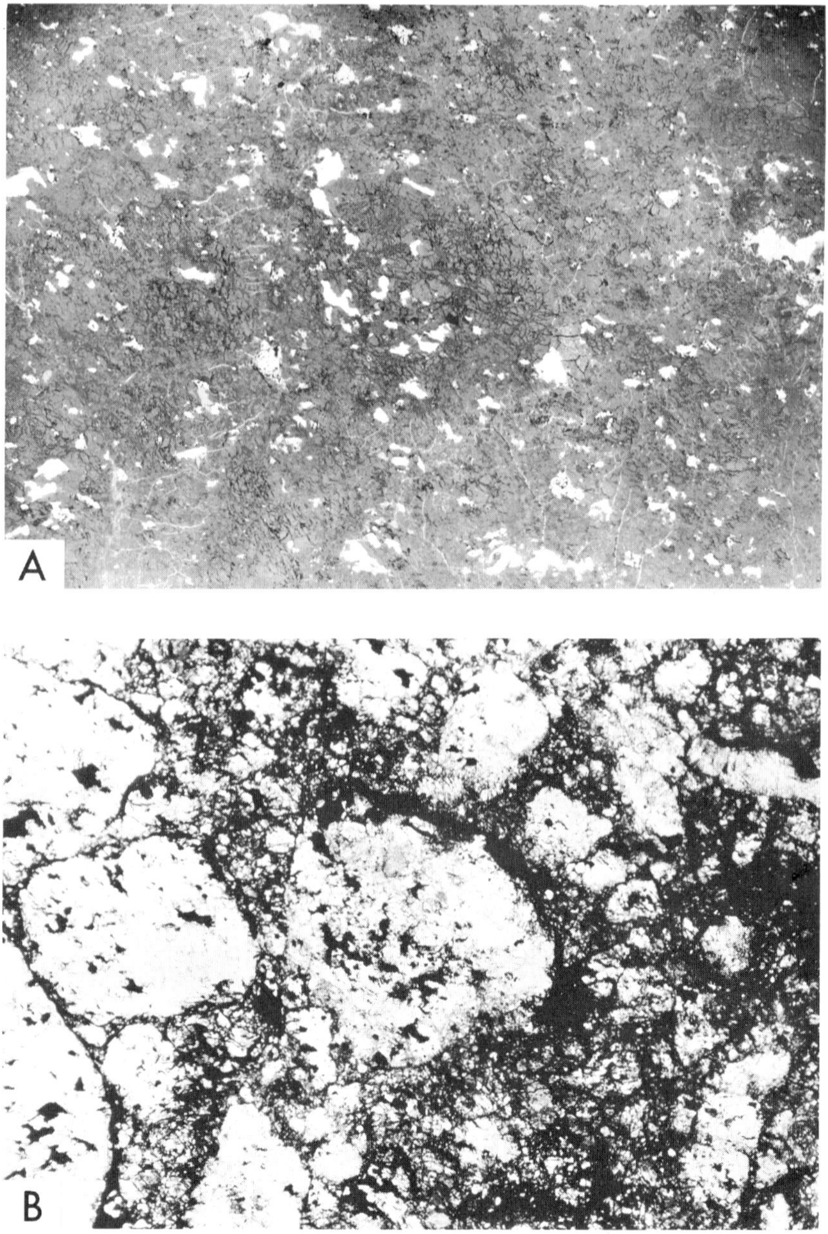

Figure 57 — Microscopic overview of the Paranaíba, Mato Grosso, Brazil L6 chondrite, consisting of angular, subrounded and rounded light areas cut and surrounded by black, glassy shock veins, resulting in a brecciated texture. A) Reflected light view showing silicates (olivine, pyroxene, maskelynite, glass) (different shades of gray) and metallic nickel-iron and troilite (white), either filling minute cracks in areas that were shock-melted or occurring as isolated grains. B) Transmitted, plane polarized light view of the same area as shown in Fig. 57A, displaying the light fragments and the black, glassy shock veins. Field of view is 7.6 mm wide.

of optical and X-ray determinations, Mason, 1967c, listed a value of Fa_{23}), and bronzite, $Fs_{20.7}$ (Table 20), with subordinate amounts of oligoclase glass (maskelynite), $Ab_{83.1}An_{11.3}Or_{5.6}$ (Table 22), metallic nickel-iron and troilite, and accessory amounts of chromite, $Cm_{78.9}Uv_{7.3}Pc_{2.9}Sp_{11.0}$ (Table 23). The dark areas, representing highly shock-transformed and shock-melted light material, are essentially made up of glasses of variable compositions, droplets and veins of metallic nickel-iron (often as an eutectic intergrowth) as well as weakly anisotropic to perfectly isotropic phases of olivine and orthopyroxene composition (Table 27). These isotropic phases are often colorless in transmitted light, but sometimes are light to dark brown.

Classification — Mineral and bulk compositions of the light portions of the meteorite are consistent with L-group classification. Specifically, olivine and orthopyroxene (Tables 19, 20; Fig. 55), plagioclase (maskelynite; Table 22), and chromite (Table 23) compositions agree well with those for L-group chondrites (Table 4). L-group classification is also indicated by the bulk chemical analysis (Table 10), particularly the ratios Fe°/Ni (5.05), Fe/SiO_2 (0.52) and Fe°/Fe (0.31), as well as the contents of total Fe (20.86%) and total metal (7.92%) (Table 4). The uniform and homogeneous compositions of the ferromagnesian minerals, the scarcity of chondrules, and the highly recrystallized texture of the matrix suggest that Paranaíba belongs to the petrologic type 6 of Van Schmus and Wood (1967) (Table 5).

Available material — 1) United States National Museum (Smithsonian Institution), Washington, D.C. (3 sp, 564.5 g); 2) Instituto de Geociências, São Paulo, Brazil (1 sp, 500 g); 3) American Museum of Natural History, New York, U.S.A. (1 sp, 53.0 g); 4) Istituto di Mineralogia, Modena, Italy (50 g); 5) Institute of Meteoritics, Albuquerque U.S.A. (5 sp, 45.3 g).

Patrimônio (19°32'S; 48°34'W)

Synonyms — Conceição Aparecida, Morro Cavado.

Fall circumstances — The Patrimônio, State of Minas Gerais, meteorite fell in the morning (9.50 hr) of August 6, 1950, at Fazenda do Morro Cavado in the small village of Patrimônio, about 18 km from Conceição Aparecida. Sound and light effects were noticed, with the body apparently traveling from the SW to the NE. At least 20 stones fell, and pieces ranging in weight from 600 to 1800 g were reported to have been collected by local residents (Fig. 58). The meteorite is listed in Leonard (1956), Keil (1960), Hey (1966) and Hutchison et al. (1977), and an abstract dealing with its petrography and chemistry was given by Santos (1972). A detailed description of the meteorite was recently published by Gomes et al. (1978a).

Texture — The meteorite shows a poorly developed chondritic texture (Fig. 59), with only a few chondrules showing discernible boundaries with the matrix. They vary in shape from round to elongate and range in apparent size from 0.3 up to 1.8 mm (mean 1.2 mm). Their internal textures vary considerably, and barred, poikilitic, porphyritic, and granular varieties were observed. Finely-crystalline (formerly glassy) material occurs as a very rare constituent of chondrules and of the matrix.

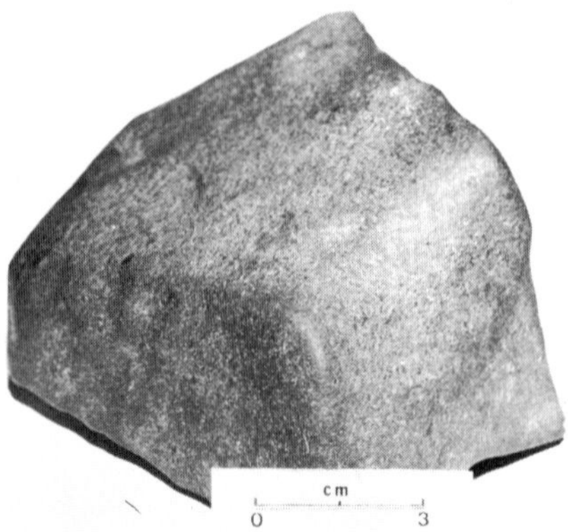

Figure 58 — Specimen of the Patrimônio, Minas Gerais, Brazil L6 chondrite weighing 1,769 g, covered by black fusion crust.

Mineralogy — Microscopic and electron microprobe investigations (Gomes et al., 1978a) indicate that the meteorite consists of olivine, $Fa_{24.8}$ (Table 19; Fa_{25}, cf., Mason, 1963, as determined by optical and X-ray diffraction methods), bronzite, $Fs_{20.7}$ (Table 20), and plagioclase, $Ab_{82.1}An_{11.3}Or_{6.6}$ (Table 22), with metallic nickel-iron and troilite occurring in minor amounts. Accessory minerals are high-Ca pyroxene (diopside), $En_{47.6}Fs_{8.5}Wo_{43.9}$ (Table 21), chromite, $Cm_{78.8}Uv_{8.2}Pc_{1.3}Sp_{11.6}$ (Table 23), and whitlockite.

Classification — Compositions of olivine, pyroxene (Tables 19, 20; Fig. 55), and chromite (Table 23) compare well with values reported for L-group ordinary chondrites (Table 4). Bulk chemical composition (Table 10) is also consistent with L-group classification, as indicated by the ratios Fe/SiO_2 (0.54), Fe°/Fe (0.24) and Fe°/Ni (4.45) and contents of total Fe (21.51%) and total metallic nickel-iron (6.25%) (Table 4). Textural and mineralogical evidence

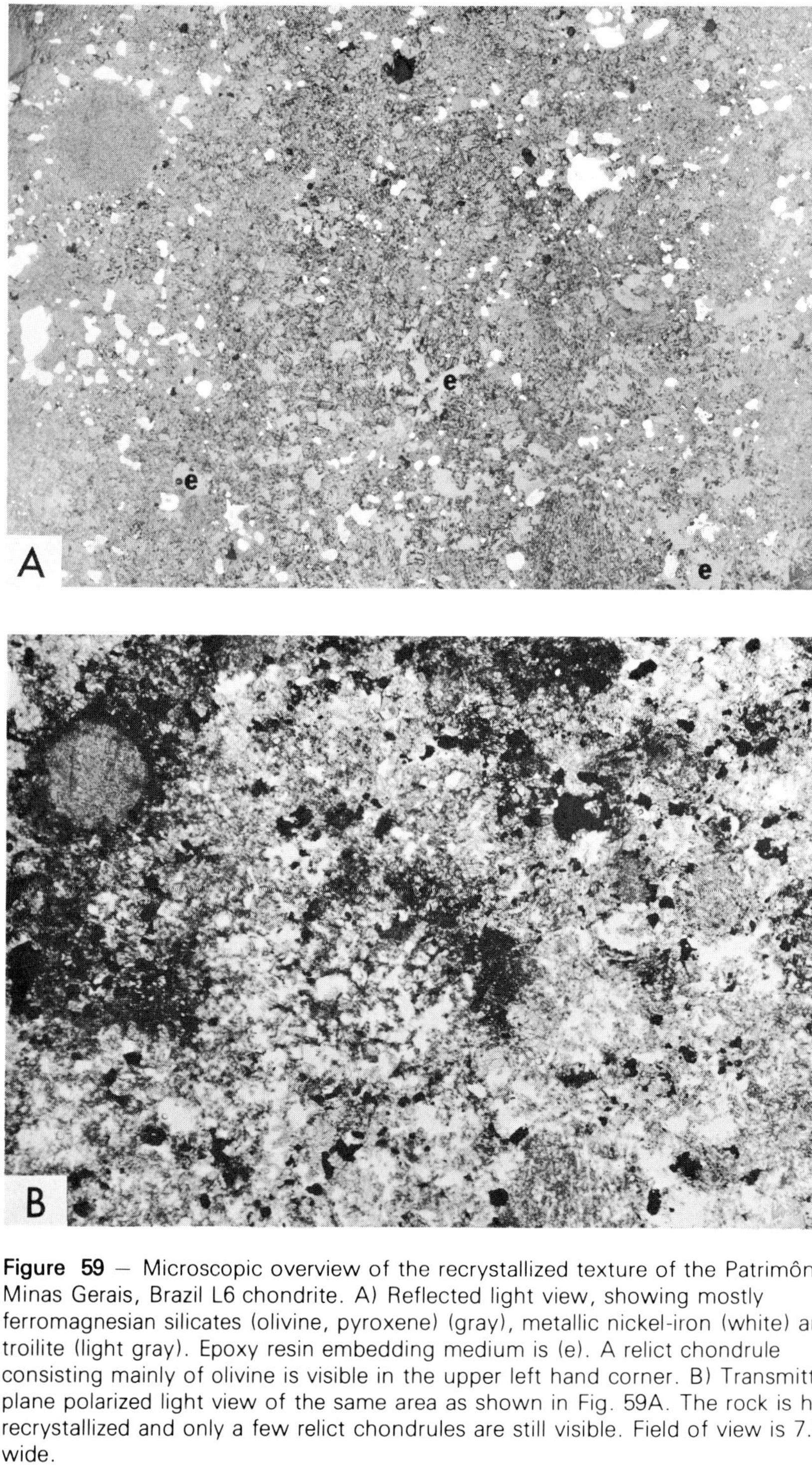

Figure 59 — Microscopic overview of the recrystallized texture of the Patrimônio, Minas Gerais, Brazil L6 chondrite. A) Reflected light view, showing mostly ferromagnesian silicates (olivine, pyroxene) (gray), metallic nickel-iron (white) and troilite (light gray). Epoxy resin embedding medium is (e). A relict chondrule consisting mainly of olivine is visible in the upper left hand corner. B) Transmitted, plane polarized light view of the same area as shown in Fig. 59A. The rock is highly recrystallized and only a few relict chondrules are still visible. Field of view is 7.6 mm wide.

(i.e., the uniform and homogeneous compositions of olivine and orthopyroxene, the only poorly defined chondritic texture, the highly recrystallized matrix, and the presence of clear interstitial plagioclase) suggests that the meteorite belongs to the petrological type 6 of Van Schmus and Wood (1967) (Table 5).

Available material — 1) Museu Nacional, Rio de Janeiro, Brazil (1 sp, 1,769 g; Fig. 58); 2) Instituto de Geociências, São Paulo, Brazil (2 sp, 54.7 g); 3) United States National Museum (Smithsonian Institution), Washington, D.C. (2 sp, 21.5 g); 4) Istituto di Mineralogia, Modena, Italy (10 g); 5) Institute of Meteoritics, Albuquerque, U.S.A. (1 sp, 1.8 g).

Putinga (29°02'S; 53°03'W)

Synonym — Encantada.

Fall circumstances — The Putinga, State of Rio Grande do Sul, meteorite fell in the afternoon (4.30 hr) of August 16, 1937, about 2 km from the locality of Putinga. The following day, two masses weighing 57 and 45 kg (Fig. 60), respectively, as well as smaller pieces up to 10 kg were recovered by local residents. The stones were reported to have been buried at depths of up to 2-3 m. The meteorite is listed in Hey (1966) and Hutchison et al. (1977), and a brief, preliminary description of its mineralogy was provided by Roisenberg (1970). More detailed studies of the rock were published by Symes and Hutchison (1970) and Keil et al. (1978b).

Texture — The meteorite exhibits a highly recrystallized texture (Fig. 61), with chondrules poorly defined and often intergrown with the matrix. They vary in shape from round to elongate and range in apparent size up to 7 mm (mean 0.8 mm). Their internal textures are quite variable, including excentro-radial, barred, porphyritic, poikilitic, and granular varieties. In addition to an extensive recrystallization, the meteorite also shows evidence of shock (e.g., transformation of feldspar into maskelynite; wavy extinction in ferromagnesian minerals).

Mineralogy — According to Symes and Hutchison (1970) and Keil et al. (1978b), the meteorite consists mainly of olivine, $Fa_{24.8}$ (Table 19; Fa_{24}, cf., Mason, 1967c, and also Symes and Hutchison, 1970), low-Ca pyroxene (bronzite), $Fs_{21.3}$ (Table 20; $Fs_{21.2}$, cf., Symes and Hutchison, 1970) and metallic nickel-iron (kamacite, taenite, plessite), with minor amounts of troilite and plagioclase. Keil et al. (1978b) observed that all plagioclase of the meteorite has been transformed into isotropic to weakly anisotropic maskelynite by solid-state shock transformation. This shock transformation must have taken place after recrystallization (i.e.,

Figure 60 — One specimen of the Putinga, Rio Grande do Sul, Brazil L6 chondrite weighing approximately 45 kg. The stone is partly covered by a black fusion crust and shows regmaglypts in places. Coin is about 25 mm in diameter.

after large, clear, crystalline plagioclase crystals had formed). The maskelynite is oligoclase in composition, $Ab_{81.0}An_{12.4}Or_{6.6}$ (Table 22) (Symes and Hutchison, 1970, listed $Ab_{82.4}An_{10.2}Or_{7.4}$ for the plagioclase composition), and is stoichiometric. Accessory phases (Keil et al., 1978b) include chromite, $Cm_{79.0}Uv_{8.2}Pc_{1.8}Sp_{11.0}$ (Table 23), and whitlockite.

Classification — Previous assignment of Putinga to the L-group chondrites (Symes and Hutchison, 1970) has been confirmed by Keil et al. (1978b), on the basis of mineral compositions, notably of olivine and pyroxene (Tables 19, 20; Fig. 55), and a new bulk chemical analysis. Plagioclase and chromite compositions (Tables 22, 23) are only approximately consistent with L-group classification (Table 4). Bulk chemical parameters (Table 10) supporting L-group classification are the ratios Fe°/Ni (5.24; 5.50, cf., Symes and Hutchison, 1970), Fe/SiO_2 (0.58; 0.58, cf., Symes and Hutchison, 1970) and Fe°/Fe (0.27; 0.28, cf., Symes and Hutchison, 1970) as well as the contents of total Fe (22.42%; 23.3%, cf., Symes and Hutchison, 1970) and total metallic nickel-iron (7.25%; 7.86%, cf., Symes and Hutchison, 1970) (Table 4). The highly recrystallized texture, with only poorly defined chondrules, and the uniform and homogeneous compositions of olivine and pyroxene, place the meteorite into petrologic type 6 of Van Schmus and Wood (1967) (Table 5), in agreement with the earlier classification by Symes and Hutchison (1970).

Available material — 1) Museu Luiz Englert, Universidade Federal do Rio Grande do Sul, Porto Alegre, Brazil (1 sp, 45 kg;

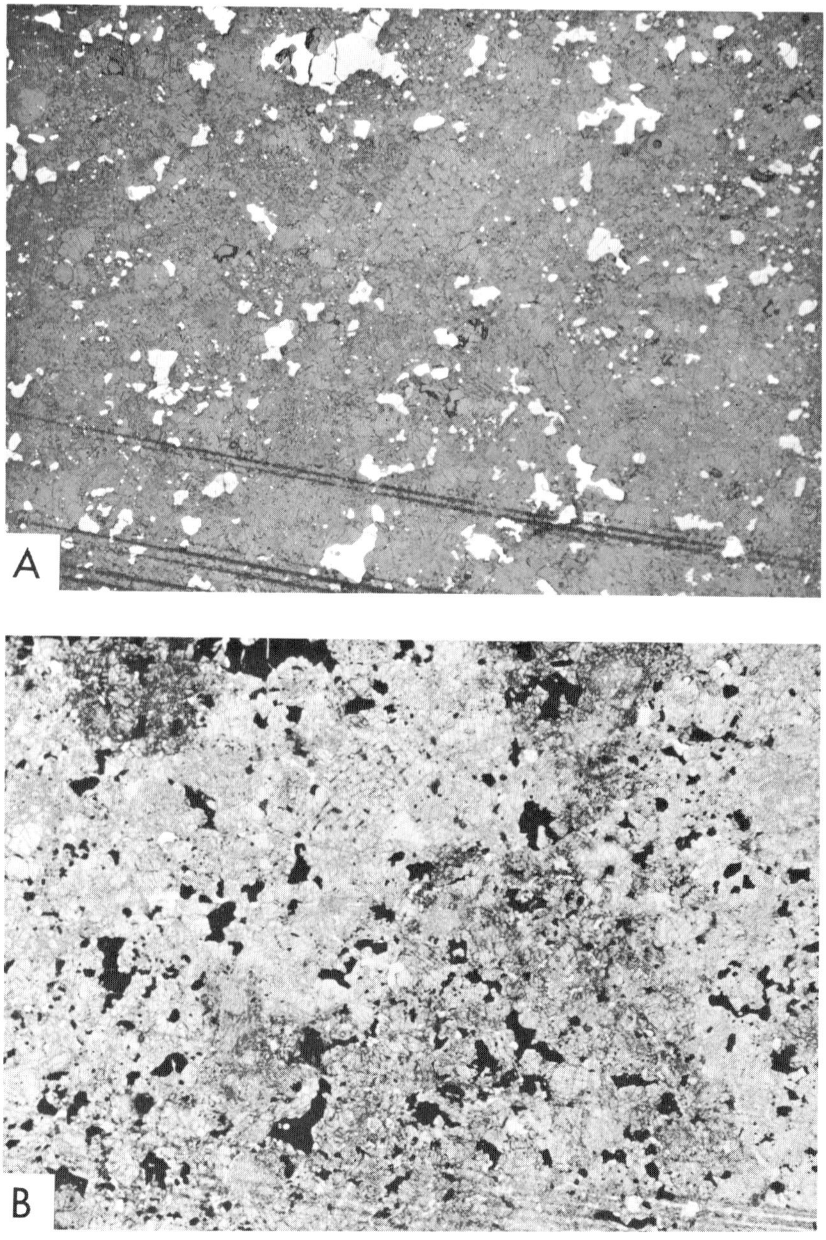

Figure 61 — Microscopic overview of the Putinga, Rio Grande do Sul, Brazil L6 chondrite. A) Reflected light view, showing mostly ferromagnesian silicates (olivine, pyroxene) (gray), metallic nickel-iron (white), troilite (tan), and only poorly discernible chondrules. B) Transmitted, plane polarized light view of the same area as shown in Fig. 61A, illustrating the highly recrystallized texture with only a few relict chondrules barely recognizable. Field of view is 7.6 mm wide.

Fig. 60); 2) United States National Museum (Smithsonian Institution), Washington, D.C. (3 sp, 967.9 g); 3) Vatican Observatory, Rome, Italy (284 g; cf., Hey, 1966); 4) American Museum of Natural History, New York (1 sp, 254.3 g); 5) Instituto de Geociências, São Paulo, Brazil (3 sp, 103 g); 6) Istituto di Mineralogia, Modena, Italy (30 g); 7) British Museum (Natural History), London, England (1 sp, 27 g); 8) Institute of Meteoritics, Albuquerque, U.S.A. (2 sp, 14.6 g); 9) Center for Meteorite Studies, Tempe, U.S.A. (1 sp, 12.3 g).

Rio Negro (26°6'S; 49°48'W)

Fall circumstances – The Rio Negro, State of Paraná, meteorite fell in the evening (8.30 hr) of September 22, 1934, at Voltagem, apparently a small farm near Rio Negro. Sound effects were noticed by local residents and material weighing slightly over 1,000 g was reported to have been recovered by Mr. Ayres Rouen, a former Rio Negro mayor. The meteorite is listed in Hey (1966) and Hutchison et al. (1977), and a preliminary description, including partial chemical analysis, was published by Gatterer and Junkes (1940). A detailed investigation of the meteorite was recently carried out by Fodor et al. (1977).

Texture – The meteorite exhibits a well-developed chondritic texture (Fig. 62), with chondrules varying in size and internal textures. Barred, radiating, and microporphyritic varieties were observed and chondrule margins are clearly discernible from the matrix, a fine-grained material that is glassy in places. In hand specimen, it was also noted that the meteorite contains a dark-colored lithic fragment, about 5 mm in size (Fig. 62).

Mineralogy – According to Fodor et al. (1977), the meteorite (host) is essentially made up of olivine and twinned, low-Ca pyroxene, both minerals varying substantially in composition (Table 28; Figs. 63, 64), and dark-colored or dusty igneous glass, which in some cases resembles feldspar in composition (Table 29). The meteorite also contains minor amounts of metallic nickel-iron (kamacite and taenite), troilite, and high- and low-Al_2O_3 chromite (Table 28). Accessory phases include high-Ca pyroxene (Fig. 64) and nepheline (Table 28). The lithic fragment has olivine and low-Ca pyroxene showing an even more variable composition than do those phases in the host (Table 30; Figs. 63, 64), abundant Fe-Mg silicates, resembling serpentine-like phases (Table 30), and glass of heterogeneous composition, with some grains being calcic, others sodic (Table 29). High-Ca-high-Al_2O_3 pyroxene and low-Al_2O_3 chromite (Table 30) occur in accessory amounts.

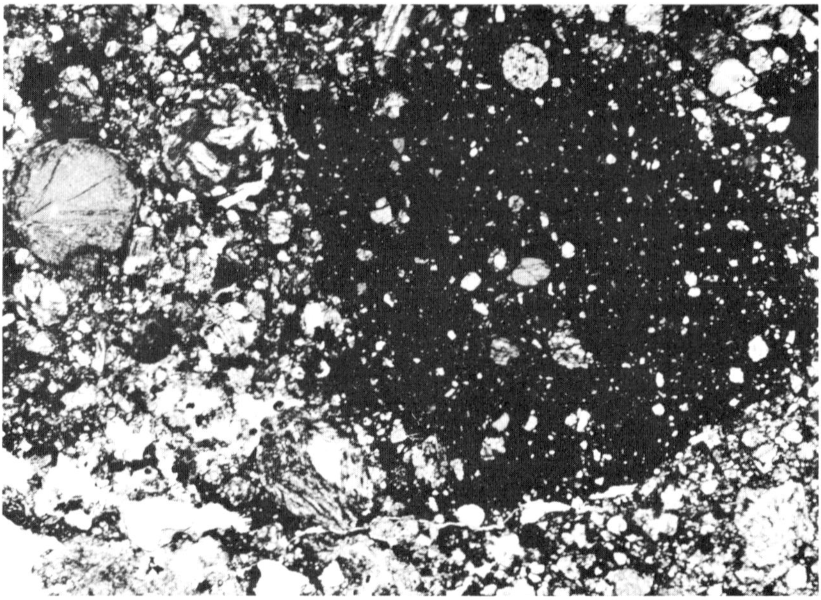

Figure 62 — Microscopic overview of the Rio Negro, Paraná, Brazil L3-4 chondrite, showing the pronounced chondritic texture and a subrounded, dark-colored lithic fragment. Transmitted, plane polarized light. Field of view is 7.6 mm wide.

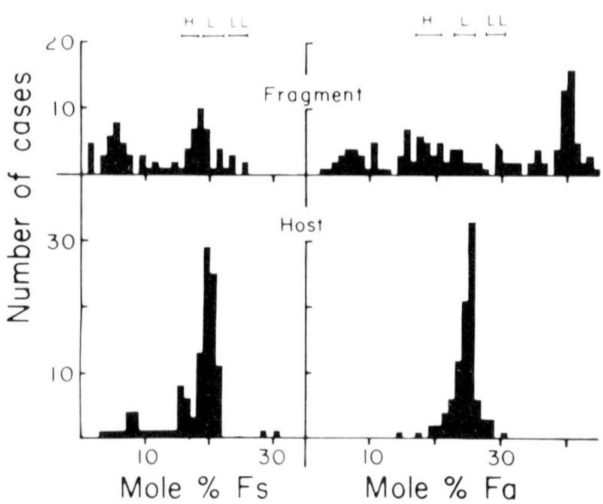

Figure 63 — Histograms illustrating the compositions of low-Ca pyroxene and olivine in the Rio Negro, Paraná, Brazil L 3-4 chondrite and a lithic fragment therein, expressed as mole % endmembers ferrosilite (Fs; $FeSiO_3$) and fayalite (Fa; Fe_2SiO_4) (after Fodor et al., 1977). Por comparison, the ranges for average olivine and pyroxene compositions in H-, L-, and LL-group chondrites are shown (after Keil and Fredriksson, 1964; as modified by Fodor et al., 1976).

Classification — Although variable in composition, olivine and low-Ca pyroxene in the host chondrite cluster mainly in the L-group compositional range (Fig. 63). Additional support for a L-group classification of Rio Negro stems from the bulk

Table 28. Representative analyses of silicate and oxide minerals in the Rio Negro host chondrite (after Fodor et al., 1977)

	Olivine			Low-Ca pyroxene			Chromite		Nepheline
SiO_2	39.1	38.1	38.8	57.0	56.0	56.0	0.49	0.44	41.8
TiO_2	0.04	0.02	0.03	0.19	0.24	0.16	2.1	0.95	n.d.
Al_2O_3	<0.01	<0.01	<0.01	0.45	0.17	0.31	4.7	22.4	31.4
Cr_2O_3	0.02	0.20	0.02	0.23	0.41	0.17	56.7	40.3	n.d.
V_2O_3	n.d.	n.d.	n.d.	n.d.	n.d.	n.d.	0.74	0.26	n.d.
FeO	21.1	23.0	23.4	6.1	11.2	13.8	30.6	25.6	0.81
MnO	0.45	0.46	0.47	0.29	0.50	0.52	0.93	0.56	n.d.
MgO	39.4	38.8	37.1	36.3	30.4	29.3	2.8	7.9	0.80
CaO	0.01	0.02	0.02	0.33	1.6	0.78	0.07	0.09	2.0
Na_2O	<0.01	<0.01	<0.01	0.05	0.10	0.01	n.d.	n.d.	16.8
K_2O	n.d.	n.d.	n.d.	n.d.	n.d.	n.d.	n.d.	n.d.	5.6
TOTAL	100.12	100.60	99.84	100.94	100.62	101.05	99.15*	98.65*	99.21
	Fo 76.9	75.0	73.9	En 90.8	80.4	77.9	Cm 78.3	52.9	
	Fa 23.1	25.0	26.1	Fs 8.6	16.6	20.6	Hc 0.0	7.3	
				Wo 0.6	3.0	1.5	Uv 7.3	3.6	
							Pc 4.6	0.0	
							Sp 9.8	36.2	

* Includes 0.02% and 0.15% ZnO, respectively.
n.d. = not determined.

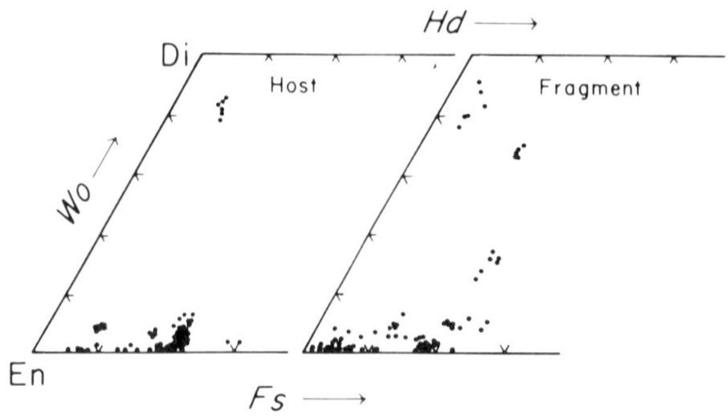

Figure 64 — Partial quadrilateral diagrams illustrating the compositions of pyroxene in the Rio Negro, Paraná, Brazil L 3-4 chondrite and its lithic fragment, expressed as mole % endmembers enstatite (En; $MgSiO_3$), ferrosilite (Fs; $FeSiO_3$), and wollastonite (Wo; $CaSiO_3$) (after Fodor et al., 1977). Di stands for diopside, $(Ca,Mg)SiO_3$, and Hd for hedenbergite, $(Ca,Fe)SiO_3$.

Table 29. Representative analyses of glass in Rio Negro host chondrite and the lithic fragment (after Fodor et al., 1977)

	Host				Fragment		
SiO_2	57.7	57.6	64.0	67.2	45.1	52.3	61.7
Al_2O_3	7.5	21.3	20.7	19.5	31.9	26.5	20.1
FeO	5.0	2.5	1.8	0.98	1.6	2.5	2.0
MgO	10.2	5.2	3.0	1.8	1.4	4.3	0.25
CaO	13.5	8.9	2.8	1.8	15.7	8.3	2.9
Na_2O	4.4	5.5	8.5	9.7	3.3	4.5	9.4
K_2O	0.55	0.86	0.94	0.90	0.24	0.60	1.2
TOTAL	98.65	101.86	101.74	101.88	99.24*	99.00	97.55

* From a poikilitic fragment within the lithic fragment.

rock composition (Table 10), specifically the ratios Fe/SiO_2 (0.54) and Fe^0/Fe (0.31), as well as the contents of total Fe (21.40%) and total metallic nickel-iron (7.91%) (Table 4). The variability in the composition of olivine and pyroxene, the well-defined chondritic texture and the presence of igneous glass suggest that Rio Negro belongs to the petrologic type 3 transitional to type 4 of Van Schmus and Wood (1967) (Table 5), in agreement with the earlier classification of the meteorite made by these authors.

Origin of the lithic fragment — From the mineral compositions (Table 30; Figs. 63, 64) and textural features, it is evident that the fragment did not form by shock darkening of the host and, thus, it must represent a xenolith. Texture, bulk and mineral compositions of the fragment resemble those of unequilibrated

Table 30. Representative analyses of silicate and oxide minerals in a lithic fragment from the Rio Negro chondrite (after Fodor et al., 1977)

	Olivine			Low-Ca pyroxene		High-Ca pyroxene (range)				Chromite	Fe-Mg silicate grains	
SiO_2	39.1	38.7	36.5	58.2	55.6	52.0	47.6	48.0	44.3	0.77	48.5	49.5
TiO_2	0.02	0.02	0.04	0.04	0.10	0.60	1.1	0.30	1.0	1.8	0.04	0.01
Al_2O_3	<0.01	<0.01	0.03	0.33	0.62	3.1	8.9	12.6	15.5	1.7	0.28	0.21
Cr_2O_3	0.09	0.08	0.08	0.38	0.68	0.96	0.92	0.08	0.60	58.5	0.43	0.43
V_2O_3	n.d.	n.d.	n.d.	n.d.	n.d.	n.d.	n.d.	n.d.	n.d.	0.83	n.d.	n.d.
FeO	14.5	20.5	33.1	3.9	12.2	12.9	8.8	2.7	2.6	31.4	22.8	23.0
MnO	0.11	0.24	0.39	0.25	0.43	0.31	0.26	0.02	0.02	0.57	0.35	0.38
MgO	44.9	40.8	30.1	36.5	27.7	22.3	16.6	17.9	16.2	2.2	24.8	25.4
CaO	0.04	0.13	0.03	0.36	2.5	7.6	15.5	18.3	19.9	0.37	3.1	1.5
Na_2O	<0.01	<0.01	<0.01	0.06	0.04	0.01	0.01	<0.01	0.02	n.d.	0.22	0.19
K_2O	n.d.	n.d.	n.d.	n.d.	n.d.	n.d.	n.d.	n.d.	n.d.	n.d.	0.05	0.03
P_2O_5	n.d.	n.d.	n.d.	n.d.	n.d.	n.d.	n.d.	n.d.	n.d.	n.d.	0.07	0.03
TOTAL	98.76	100.47	100.27	100.02	98.87	99.78	99.69	99.90	100.14	98.23*	100.64	100.68
	Fo 84.7	78.0	61.8	En 93.7	76.2	64.7	50.8	54.9	50.7	Cm 80.8		
	Fa 15.3	22.0	38.2	Fs 5.6	18.8	20.1	15.1	4.7	4.6	Uv 7.6		
				Wo 0.7	5.0	15.2	34.1	40.4	44.7	Pc 8.0		
										Sp 3.6		

* Includes 0.09% ZnO.
n.d. = not determined.

ordinary chondrites and carbonaceous chondrites of type III, although some properties (particularly the preponderance of olivine compositions near Fa_{40}, Fig. 63) suggest a closer relationship to type III carbonaceous chondrites. The source area of the fragment (coexisting material on the same parent body or residue of an impacting projectile) is a matter of speculation, but it is clear that the fragment was implanted into the meteorite host by impact processes, and that it once was part of the regolith that covered the Rio Negro parent body.

Available material – 1) Vatican Observatory, Rome, Italy (694 g; cf., Hey, 1966); 2) United States National Museum (Smithsonian Institution), Washington, D.C. (1 sp, 55.0 g); 3) British Museum (Natural History), London, England (1 sp, 50 g); 4) Max-Planck-Institut für Chemie, Mainz, Federal Republic of Germany (1 sp, 39.9 g); 5) Naturhistorisches Museum Wien, Vienna, Austria (1 sp, 10 g).

Santa Bárbara (29°12′S; 51°52′W)

Synonym – Rio Grande do Sul.

Fall circumstances – The Santa Bárbara, State of Rio Grande do Sul, meteorite fell at 12.00 noon on either September 26, 1873 (most probable date), or on an unknown day in October, 1873. The fall was observed by some witnesses (local farmers) in a cloudless sky, with the meteorite traveling from the N to the S. Sound effects (three loud detonations followed by machine gun fire) were noticed and one single stone (400 g) was recovered (Derby, 1888a, b). The meteorite is also listed in Oliveira (1931), Vidal (1936) and Hey (1966), and a detailed description was given by Berkley et al. (1978b).

Texture – The meteorite shows a well-developed chondritic texture (Fig. 65), with the chondrules readily discernible from the surrounding matrix, a microcrystalline, granular, and transparent material. In some areas of the matrix, a turbid, microcrystalline "glass" is also found. Chondrules vary considerably in texture and include porphyritic, poikilitic, barred, and excentro-radial varieties.

Mineralogy – According to Berkley et al. (1978b), the meteorite consists mainly of olivine, $Fa_{24.6}$ (Table 19; Fa_{25}, cf., Mason, 1963, on the basis of optical and X-ray determinations), low-Ca pyroxene (bronzite), $Fs_{20.6}$ (Table 20), and metallic nickel-iron, with accessory amounts of high-Ca pyroxene (diopside), $En_{51.3}Fs_{10.4}Wo_{38.3}$ (Table 21), microcrystalline plagioclase, $Ab_{84.8}An_{10.0}Or_{5.2}$ (Table 22), chromite, $Cm_{81.9}\ Uv_{5.6}Pc_{1.1}Sp_{11.4}$ (Table 23), and glassy material.

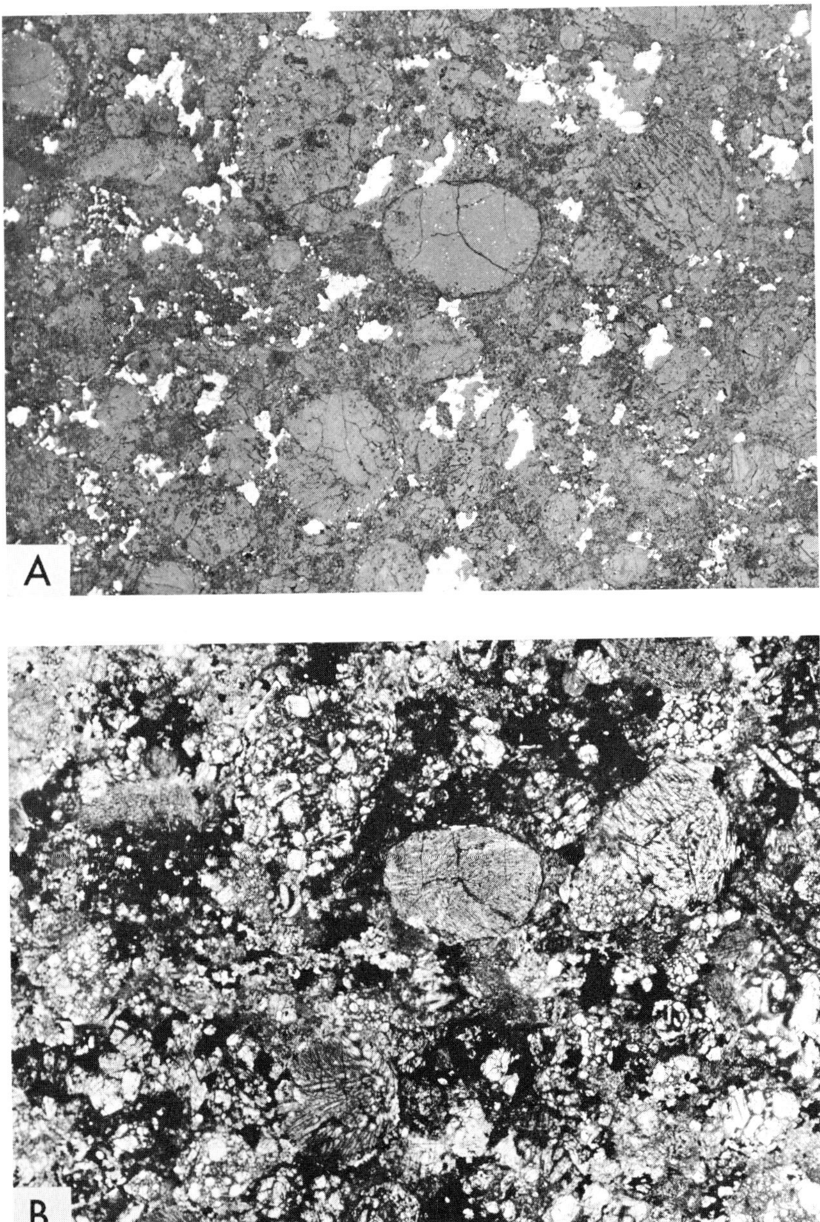

Figure 65 – Microscopic overview of the texture of the Santa Bárbara, Rio Grande do Sul, Brazil L4 chondrite. A) Reflected light view, showing mostly ferromagnesian silicates (olivine, pyroxene) (gray), metallic nickel-iron and troilite (white), and well-developed, clearly discernible chondrules. B) Transmitted, plane polarized light view of the same are as shown in Fig. 65A, illustrating the well-developed chondritic texture of the meteorite and various textural chondrule types. Field of view is 7.6 mm wide.

Classification — Chemical composition of the constituent phases, notably olivine, low-Ca pyroxene (Tables 19, 20; Fig. 55), and plagioclase (Table 22), indicate that Santa Bárbara belongs to the L-group chondrites (Table 4). Textural evidence, e.g., the abundance of twinned low-Ca clinopyroxene, the well-preserved chondritic texture, the presence of turbid glass, and the occurrence of rare and microcrystalline plagioclase, suggest that the meteorite is only mildly recrystallized and belongs to the petrologic type 4 of Van Schmus and Wood (1967) (Table 5).

Available material — 1) Field Museum of Natural History, Chicago, U.S.A. (1 sp, 41 g); 2) Museu Nacional, Rio de Janeiro, Brazil (1 sp, 34.6 g); 3) Naturhistorisches Museum Wien, Vienna, Austria (1 sp, 2g); 3) British Museum (Natural History), London, England (1 sp, 1.5 g); 4) Muséum National d'Histoire Naturelle, Paris, France (1 sp, 1.4 g).

7.2.3. LL-group chondrite

Parambu (6° 14'S; 40° 42'W)

Fall circumstances — The Parambu, State of Ceará, meteorite fell in the evening (approximately 7.00 hr) of July 24, 1967, at Fazenda Novo Exu, a ranch near the locality of Parambu. A fireball was seen traveling from SW to NE and the strewnfield measured about 85,000 m^2. After detonations, a large number of stones fell and more than twenty-seven specimens, the largest one weighing 594.6 g, were reported to have been recovered by local people. The fall was first reported by Barreto et al. (1973) and the meteorite is listed in Hutchison et al. (1977). A mineralogical-petrographic description was given by Levi-Donati and Sighinolfi (1974) and a bulk chemical analysis was published by Shima et al. (1974). A detailed electron microprobe study of the rock was made by Gomes et al. (in preparation).

Texture — The most conspicuous feature of the Parambu chondrite is its brecciated texture. The rock consists of a fine-grained matrix of variable grain size that contains well-developed chondrules, subrounded to rounded in shape, with complex internal textures (barred, excentro-radial, porphyritic, etc.) (Fig. 66). Most of the matrix consists of fragments of clasts and, in its grain size, is transitional to the larger clasts (Figs. 66, 67). The clasts are quite variable in color, size, shape, and texture (Fig. 67). A few fragments are dark, the color being due to the dissemination of tiny opaque phases, notably metal and troilite. Some white clasts are highly recrystallized and have clear, interstitial plagioclase grains. The texture is typical of regolith-type breccias, with the clasts ranging in shape from angular to subrounded to rounded (abraded).

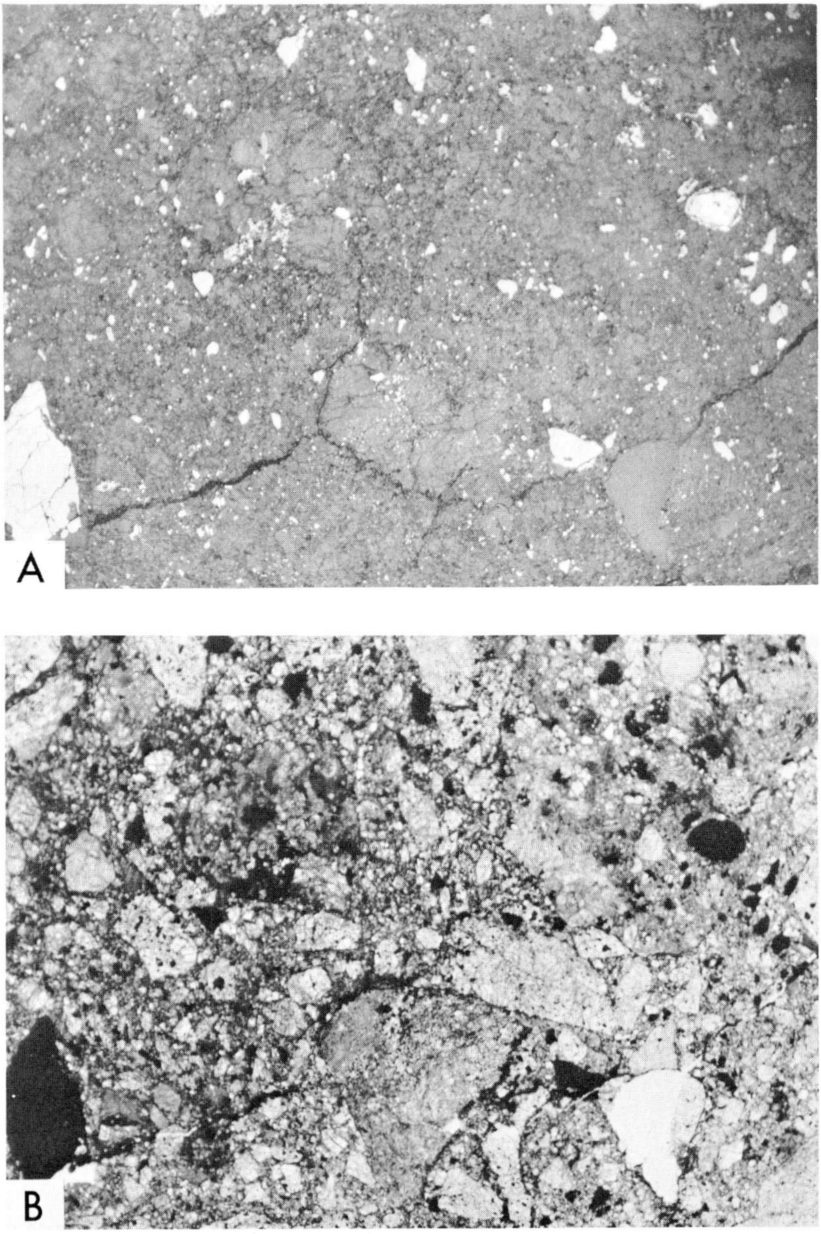

Figure 66 — Microscopic overview of the Parambu, Ceará, Brazil LL chondrite. A) Reflected light view, showing a brecciated texture and mostly ferromagnesian minerals (gray), metallic nickel-iron (white) and troilite. B) Transmitted, plane polarized light view of the same area as shown in Fig. 66A, displaying the highly brecciated texture of Parambu, consisting of angular to subrounded and rounded lithic fragments of highly variable size and texture, embedded in a finer-grained matrix. Field of view is 7.6 mm wide.

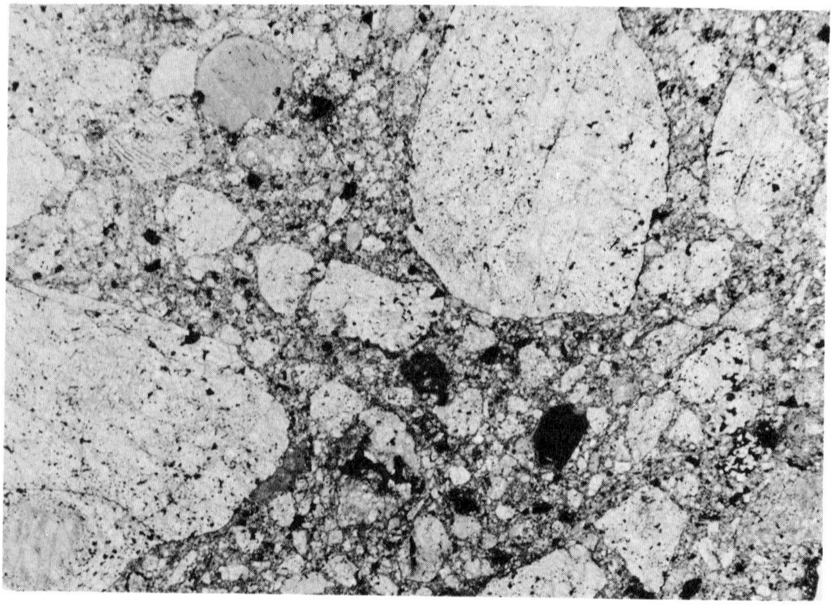

Figure 67 — Photomicrograph of the Parambu, Ceará, Brazil LL chondrite, showing the highly brecciated texture of the meteorite, consisting of angular to subrounded and rounded lithic fragments of highly variable size and texture that are embedded in a finer-grained matrix. Two chondrule-like objects are visible in the upper left hand corner. Transmitted, plane polarized light. Field of view is 7.6 mm wide.

According to Levi-Donati and Sighinolfi (1974), the lithic portion represents 91.5% of the meteorite and the opaque components the remaining 8.5%.

Mineralogy — The meteorite consists essentially of olivine, orthorhombic and monoclinic pyroxene, nickel-iron, and troilite; plagioclase and chromite are minor constituents. Rare and small grains of maskelynite, whitlockite, ilmenite, and pentlandite are also present (Levi-Donati and Sighinolfi, 1974; Gomes et al., 1978b).

Levi-Donati and Sighinolfi (1974) listed values of Fa_{27-28} and Fa_{28} for the olivine composition of Parambu on the basis of refractive index and X-ray measurements, respectively. Also, the pyroxene and plagioclase compositions, as determined from the refractive indices, were given by these authors as Fs_{22-23} and An_{10}, respectively. Probe data (Gomes et al., in preparation) for olivine and low-Ca pyroxene are listed in Tables 31 and 32 and are plotted in Fig. 68. The histograms represent cumulative data obtained from four sections and include matrix and chondrules (upper histogram) and lithic clasts (average of nine different fragments per section). In spite of some unusual compositions (indicated in Tables 31, 32), the results are quite consistent for

Table 31. Average compositions of olivine in individual lithic fragments from the Parambu, Ceará, chondrite (in wt. %). A, B and C designate different sections (after Gomes et al., in preparation)

	A.II	A.VII	A.X	B VII	B.X	C.VII	Typical Composition*	Matrix
SiO_2	38.4	38.2	38.7	38.4	38.7	38.3	37.9	38.1
TiO_2	<0.01	0.02	0.04	0.11	0.02	<0.01	<0.01	<0.01
Al_2O_3	<0.01	<0.01	<0.01	<0.01	<0.01	<0.01	<0.01	<0.01
FeO	26.6	28.0	24.4	25.6	24.8	25.6	28.6	27.8
MnO	0.33	0.34	0.33	0.35	0.32	0.36	0.31	0.33
MgO	35.5	34.2	37.3	35.7	36.6	36.6	34.0	34.3
CaO	<0.01	0.03	<0.01	0.03	<0.01	<0.01	0.02	<0.01
TOTAL	100.83	100.79	100.77	100.19	100.44	100.86	100.83	100.53
Fo	70.4	68.5	73.2	71.4	72.5	71.8	67.9	68.7
Fa	29.6	31.5	26.8	28.6	27.5	28.2	32.1	31.3

* Typical composition of olivine in a selected fragment of the most common type.

Table 32. Average compositions of low-Ca pyroxene in individual lithic fragments from the Parambu, Ceará, chondrite (in wt. %). A, B and C designate different sections (after Gomes et al., in preparation)

	A.II	A.VII	A.VIII	A.X	B.VII	B.X	C.VII	Typical Composition*	Matrix
SiO_2	55.0	55.6	56.6	55.4	54.1	54.2	54.5	55.0	55.5
TiO_2	0.22	0.20	0.05	0.02	0.09	0.12	0.15	0.15	0.16
Al_2O_3	0.14	0.10	0.20	0.11	0.10	0.05	0.09	0.08	0.12
FeO	16.0	17.1	11.6	15.1	15.5	16.1	16.6	16.8	16.9
MnO	0.47	0.33	0.26	0.45	0.29	0.35	0.36	0.39	0.36
MgO	27.1	27.0	30.4	28.7	27.4	27.2	26.7	26.6	27.1
CaO	0.81	0.93	0.57	0.32	0.81	0.76	0.92	0.83	0.86
Na_2O	<0.01	0.03	0.13	0.02	<0.01	<0.01	<0.01	<0.01	0.03
TOTAL	99.74	101.29	99.81	100.12	98.29	98.78	99.32	99.85	101.03
En	74.0	72.5	81.5	76.8	74.7	74.0	72.8	72.6	72.9
Fs	24.4	25.7	17.4	22.6	23.7	24.6	25.4	25.7	25.5
Wo	1.6	1.8	1.1	0.6	1.6	1.5	1.8	1.6	1.7

* Typical composition of low-Ca pyroxene in a selected fragment of the most common type.

both matrix and fragments and are different from those given by Levi-Donati and Sighinolfi (1974): olivine compositions cluster in the ranges of Fa_{31-32} and pyroxene compositions in the ranges of Fs_{25-26}. Typical compositions for high-Ca pyroxene, plagioclase, and chromite are $En_{46.4}Fs_{10.6}Wo_{43.0}$ (Table 33), $Ab_{80.2}An_{10.6}Or_{9.3}$ (Table 34), and $Cm_{76.6}Hc_{2.0}Uv_{11.6}Sp_{9.8}$ (Table 35), respectively.

Table 33. Average compositions of high-Ca pyroxene in individual lithic fragments from the Parambu, Ceará, chondrite (in wt. %). A, B and C designate different sections (after Gomes et al., in preparation).

	A.II	A.VII	A.X	B.X	C.VII	Typical Composition*	Matrix
SiO_2	54.1	54.0	54.8	52.9	53.2	54.2	54.6
TiO_2	0.38	0.39	0.27	0.42	0.38	0.40	0.40
Al_2O_3	0.44	0.44	0.68	0.49	0.44	0.44	0.50
FeO	5.89	6.70	5.73	5.77	6.36	6.49	6.62
MnO	0.22	0.19	0.21	0.13	0.22	0.25	0.17
MgO	16.0	16.3	15.9	15.9	15.9	15.9	16.3
CaO	21.0	21.7	20.6	21.8	21.2	20.5	21.5
Na_2O	0.48	0.52	0.66	0.49	0.52	0.48	0.53
TOTAL	98.51	100.24	98.85	97.90	98.22	98.66	100.62
En	46.5	45.7	46.9	45.6	45.8	46.4	46.0
Fs	9.6	10.5	9.5	9.3	10.3	10.6	10.5
Wo	43.9	43.7	43.7	45.1	43.9	43.0	43.6

* Typical composition of high-Ca pyroxene in a selected fragment of the most common type.

Classification – Levi-Donati and Sighinolfi (1974), on the basis of olivine and pyroxene compositions determined by refractive index and X-ray diffraction measurements, concluded that Parambu is an L-group chondrite. However, a bulk chemical analysis by Shima et al., (1974) (Table 10) clearly shows that the meteorite belongs to the LL-group, particularly on the basis of the ratios Fe/SiO_2 (0.46) and Fe^0/Fe (0.025) and the very low contents of total Fe (18.95%), and total metallic nickel-iron (1.418%) (Table 4). LL-group classification is also indicated by the abundance pattern for heavy trace elements (Shima et al., 1974) (Table 36) and by our electron microprobe data for olivine and low-Ca pyroxene (Tables 31, 32; Fig. 68) (Gomes et al., in preparation). Cumulative patterns for the matrix and the lithic clasts are essentially identical and, although at the high-iron end of the LL-group, on the average clearly demonstrate LL-group classification for the rock.

Table 34. Average compositions of plagioclase in individual lithic fragments from the Parambu, Ceará, chondrite (in wt. %). A, B and C designate different sections (after Gomes et al., in preparation)

	A.II	A.X	B.VII					B.X	C.VII	Typical Composition*	Matrix
SiO_2	66.0	65.4	46.5	46.4	48.4	50.5	57.6	66.3	65.9	65.3	64.4
Al_2O_3	21.4	21.6	34.0	33.7	31.7	30.7	25.5	21.0	21.1	21.5	21.0
FeO	0.54	1.28	0.32	0.48	0.40	0.54	0.29	0.95	0.75	0.71	0.77
MgO	<0.01	0.04	0.13	0.04	0.24	0.06	0.02	0.20	0.12	0.02	0.11
CaO	2.25	2.17	16.9	16.7	14.1	13.6	8.83	2.02	2.27	2.27	2.18
Na_2O	9.14	9.08	1.61	1.96	3.04	3.67	4.68	9.49	9.14	9.50	8.63
K_2O	1.37	0.96	0.06	0.09	0.10	0.11	0.20	0.43	0.54	1.67	0.77
TOTAL	100.70	100.53	99.52	99.37	97.98	99.18	97.12	100.39	99.82	100.97	97.86
Ab	81.0	83.3	14.6	17.4	27.9	32.6	48.3	87.1	85.1	80.2	83.4
An	11.0	11.0	85.0	82.1	71.5	66.7	50.4	10.3	11.7	10.6	11.7
Or	8.0	5.8	0.4	0.5	0.6	0.6	1.3	2.6	3.3	9.3	4.9

* Typical composition of plagioclase in a selected fragment of the most common type.

Table 35. Average compositions of chromite in individual lithic fragments from the Parambu, Ceará, chondrite (in wt. %). A, B and C designate different sections (after Gomes et al., in preparation)

	A.II	A.X	B.VII	B.X	C.VII	Typical Composition*	Matrix
SiO_2	0.35	0.73	0.38	0.60	0.48	0.16	0.30
TiO_2	3.85	1.80	3.32	2.39	1.03	4.11	3.63
Al_2O_3	4.99	4.86	5.78	4.80	5.29	5.52	5.42
Cr_2O_3	55.6	59.2	54.3	57.2	58.9	52.6	54.5
V_2O_3	0.70	0.74	0.75	0.77	0.71	0.69	0.71
FeO	32.5	31.1	32.7	31.6	30.5	33.4	32.9
MnO	0.66	0.78	0.66	0.78	0.73	0.67	0.66
MgO	1.97	1.84	1.77	1.90	1.80	1.84	1.87
TOTAL	100.62	101.05	99.66	100.04	99.44	98.99	99.99
Cm	77.1	82.0	77.5	80.8	84.3	76.6	77.7
Hc	0.0	0.0	2.6	0.0	1.4	2.0	1.3
Uv	11.8	7.6	10.4	8.7	4.6	11.6	10.9
Pc	0.4	0.1	0.0	0.3	0.0	0.0	0.0
Sp	10.6	10.4	9.6	10.2	9.7	9.8	10.0

* Typical composition of chromite in a selected fragment of the most common type.

Table 36. Trace element abundances in the Parambu chondrite (after Shima et al., 1974). Values are in ppb.

W	110
Re	38
Os	620
Ir	360
Pt	740
Au	140
Hg	1330
Tl	21
Pb	350
Bi	25
Th	52
U	15

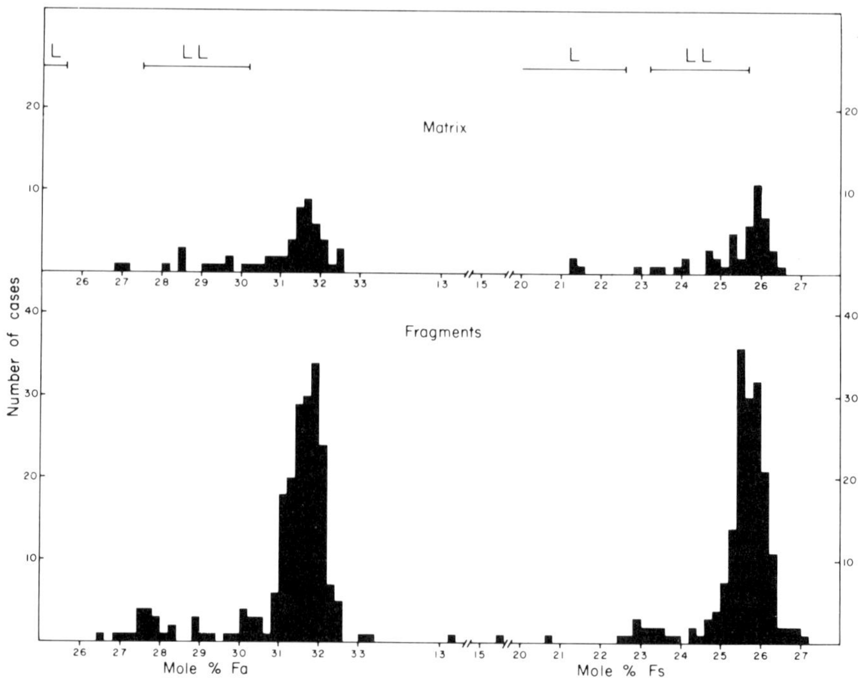

Figure 68 — Cumulative histograms for the Parambu chondrite showing the compositions, in terms of mole % endmembers, of olivine (Fa; Fe_2SiO_4) and low-Ca pyroxene (Fs; $FeSiO_3$). Data are from four sections and include matrix, lithic fragments (average of nine different data per section), and chondrules. For comparison, ranges in average compositions of olivine and pyroxene in equilibrated L- and LL-group chondrites are given (Keil and Fredriksson, 1964; as revised by Fodor et al., 1976).

Assignment of the rock into a petrologic type after Van Schmus and Wood (1967) is more difficult, however. The meteorite is a complex breccia, consisting of clasts of vastly differing textures, differing compositions and differing degrees of equilibration. The matrix, to a great portion, consists of smaller and smaller pieces of lithic clasts and it is difficult, if not impossible, to separate matrix from clasts. However, the major silicate phases in the matrix (e.g., Fig. 68) vary in composition, just as they do in some clasts. We therefore conclude that Parambu consists of a mechanical mixture of lithic clasts and matrix, each of which consists of highly-equilibrated (in the sense of Van Schmus and Wood, 1967), and unequilibrated (LL-3 in the sense of Van Schmus and Wood, 1967), as well as transitional portions. Thus, Parambu is a genomict breccia, unequilibrated overall (LL-3), but containing more equilibrated and transitional than LL-3 material (Fig. 68). We interpret this rock as a breccia formed in the regolith of its parent body. Its components were derived from unequilibrated and equilibrated rocks of LL-group composition that were mechanically mixed into what is now Parambu.

Available material — 1) Dr. A. Barreto, Divisão de Recursos Naturais (SUDENE), Recife, Brazil (1,042 g); 2) Istituto di Mineralogia, Modena, Italy (500 g); 3) Institute of Meteoritics, Albuquerque, U.S.A. (2 sp, 117.2 g); 4) Instituto de Geociências, São Paulo, Brazil (1 sp, 22.7 g); 5) Center for Meteorite Studies, Tempe, U.S.A. (1 sp, 3.5 g).

8. ACKNOWLEDGEMENTS

This book has greatly profited from contributions by many colleagues, students and friends. We are most grateful to S.E. Amaral and M.R. Arruda (Instituto de Geociências, Universidade de São Paulo, São Paulo, Brazil), F.M. Arid (Faculdade de Filosofia, Ciências e Letras de São José do Rio Preto, São José do Rio Preto, Brazil), A. Barreto and Z.F. Mello (Divisão de Recursos Naturais-SUDENE, Recife, Brazil), G. Cavarretta (Istituto di Mineralogia e Petrografia, Università di Roma, Rome, Italy), R.S. Clarke, Jr. and E.P. Henderson (Smithsonian Institution, Washington, D.C., U.S.A.), W.S. Curvello (Museu Nacional, Rio de Janeiro, Brazil), C.M. Ferreira and A.B. Oliveira (Escola de Minas, Ouro Preto, Brazil), J.E.P. Guimarães (Instituto Geológico, São Paulo, Brazil), G. Kurat (Naturhistorisches Museum, Vienna, Austria), R.L.L. Murta (Instituto de Geociências, Universidade Federal de Minas Gerais, Belo Horizonte, Brazil), L.E. Nyquist (Johnson Space Center, Houston, U.S.A.), M. Prinz (American Museum of Natural History, New York, U.S.A.), B. Riedel (Centro de Estudos Astronômicos de Minas Gerais, Belo Horizonte, Brazil), A. Roisenberg (Instituto de Geociências, Universidade Federal do Rio Grande do Sul, Porto Alegre, Brazil) and J.M.L.U. Silva (Observatório Astronômico do Colégio Estadual do Paraná, Curitiba, Brazil) for providing meteorite specimens for study.

We are greatly indebted to staff and students of the Institute of Meteoritics, University of New Mexico, Albuquerque, N.M., U.S.A., for valuable assistance. In particular, we thank G.J. Taylor and J.L. Berkley for review of the manuscript, G.H. Conrad for assistance in electron microprobe work, D.E. Lange for preparation of polished thin sections, W.L. Mansker for photographic work, G. Lux for literature researches, and M. Fillmon for typing of the manuscript. Thanks are also due to Marta J. Silva and Jaime S. Marcos of the Instituto de Geociências, Universidade de São Paulo, São Paulo, Brazil, for typing of the tables and photographing meteorite hand-specimens, respectively. Ruth Fruland (Johnson Space Center, Houston, U.S.A.) kindly carried out scanning electron microscopy of crystals in vugs of Ibitira, and E. Jarosewich (Smithsonian Institution, Washington, D.C., U.S.A.) provided unpublished bulk chemical analyses of Brazilian meteorites.

This work is supported in part by the National Aeronautics and Space Administration (Grants NGL 32-004-063 and NGL 32-004-064; K. Keil, Principal Investigator); Instituto de

Geociências da Universidade de São Paulo; and by grants from Fundação de Amparo à Pesquisa do Estado de São Paulo (Geologia 707/75, 308/77 and 405/77) and Conselho Nacional de Desenvolvimento Científico e Tecnológico (Proc. 1112.0640/77) to C.B. Gomes. The support from these agencies is gratefully achnowledged.

9. REFERENCES

ALBEE, A.L. and CHODOS, A.A. — Microprobe investigations on Apollo 11 samples. Proc. Apollo 11 Lunar Sci. Conf., pp. 135-157, 1970.

ALBEE, A.L., CHODOS, A.A., DYMEK, R.F., GANCARZ, A.J., GOLDMAN, D.S., PAPANASTASSIOU, D.A. and WASSERBURG, G.J. — Dunite from the lunar highlands: petrography, deformational history, Rb/Sr age. Lunar Sci. V, pp. 3-5, 1974.

AMARAL, S.E. — Nota preliminar sobre um meteorito caído no Sul de Mato Grosso. Bol. Soc. Bras. Geol., **11**:5-19, 1962.

ANDERS, E. — Meteorite Ages. *In*: The Solar System — IV. The Moon, Meteorites and Comets (B.M. Middlehurst and G.P. Kuiper, Eds.), pp. 402-495, University of Chicago Press, 1963.

ANDERS, E. — Origin, age and composition of meteorites. Space Scì. Rev., **3**:583-714, 1964.

ANDERS, E. — Meteorites and the early solar system. Ann. Rev. Astron. Astrophys., **9**:1-34, 1971a.

ANDERS, E. — How well do we know "cosmic" abundances? Geochim. Cosmochim. Acta, **35**:516-522, 1971b.

ANDERS, E. — Interrelations of meteorites, asteroids and comets. *In*: Physical Studies of the Minor Planets (T. Gehrels, Ed.), pp. 429-446, NASA, Washington, D.C., 1972.

ANDERS, E. — Do stony meteorites come from comets? Icarus, **24**:363-371, 1975.

ANDERS, E. — Most stony meteorites come from the asteroid belt. *In*: Asteroids: An Exploration Assessment (D. Morrison and W.C. Wells, Eds.), pp. 57-75, NASA CP-2053, U.S. Government Printing Office, Washington, D.C., 1978.

ANDERS, E. and MELLICK, P.J. — Orbital clues to the nature of meteorite parent bodies. *In*: Meteorite Research (P.M. Millman, Ed.), pp. 559-572, D. Reidel, 1969.

ANDERSEN, C.A., KEIL, K. and MASON, B. — Silicon oxynitride: a meteoritic mineral. Science, **146**:256-257, 1964.

ANGELUCCI, A. and FUNICIELLO, R. — Pioggia meteoritica a Lajeado Ipiranga, Foz de Iguaçu, Brasile, nel Dicembre 1972. Acad. Naz. Lincei, Quaderno **215**:1-19, 1975.

ARID, F.M. and RÜEGG, N.R. — Nota preliminar sobre um meteorito de Avanhandava, SP. Ciência e Cultura, **17**:128, 1965.

ARID, F.M. and RÜEGG, N.R. — Meteorito de Avanhandava, SP. An. Acad. brasil. Ciênc., **41**:155-160, 1969.

ARNOLD, J.R. — The origin of meteorites as small bodies. *In*: Isotopic and Cosmic Chemistry (H. Craig, S.L. Miller, and G.J. Wasserburg, Eds.), pp. 347-363, North Holland, Amsterdam, 1964.

ARNOLD, J.R. — The origin of meteorites as small bodies II. The model. Astrophys. J., **141**:1536-1547, 1965a.

ARNOLD, J.R. — The origin of meteorites as small bodies III. General considerations. Astrophys. J., **141**:1548-1556, 1965b.

ARRUDA, M.R. — Meteorito "Can-Can" e sua composição mineralógica. Ciência e Cultura, **14**:154, 1962.

AVANZO, P.E., LEVI-DONATI, G.R. and SIGHINOLFI, G.P. — The Marília meteorite shower — a preliminary report. Meteoritics, **8**:141-147, 1973.

BANNISTER, F.A. — Osbornite, meteoritic titanium nitride. Mineral. Mag., **26**:36-44, 1941.

BARRETO, A., MELLO, Z.F., LEVI-DONATI, G.R. and SIGHINOLFI, G.P. — The meteorite shower of Parambu, Ceará State, Brazil: mineralogy and petrology. Meteoritics, **8**:324, 1973.

BASS, M.N. — Montmorillonite and serpentine in Orgueil meteorite. Geochim. Cosmochim. Acta, **35**:139-148, 1971.

BERKLEY, J.L., TAYLOR, G.J. and KEIL, K. — Fluorescent accessory phases in the carbonaceous matrix of ureilites. Geophys. Res. Lett., **5**:1075-1078, 1978a.

BERKLEY, J.L., KEIL, K., GOMES, C.B. and CURVELLO, W.S. — Studies of Brazilian meteorites XII. Mineralogy and petrology of the Santa Bárbara, Rio Grande do Sul, chondrite. An. Acad. brasil. Ciênc., **50**:191-196, 1978b.

BERKLEY, J.L., KEIL, K., PRINZ, M. and GOMES, C.B. — The Governador Valadares nakhlite and its relationship to other nakhlites. Lunar and Planetary Sci. X, pp. 101-103, 1979.

BERKLEY, J.L., BROWN IV, H.G., KEIL, K., CARTER, N.L., MERCIER, J. C.C. and HUSS, G. — The Kenna ureilite: an ultramafic rock with evidence for igneous, metamorphic and shock origin. Geochim. Cosmochim. Acta, **40**:1429-1437, 1976.

BHANDARI, N., BHAT, S.G., LAL, D., RAJAGOPALAN, G., TAMHANE, A.S. and VENKATAVARADAN, V.S. — Super-heavy elements in extraterrestrial samples. Nature, **230**:219-224, 1971a.

BHANDARI, N., BHAT, S.G., LAL, D., RAJAGOPALAN, G., TAMHANE, A.S. and VENKATAVARADAN, V.S. — Fossil tracks in the meteorite Angra dos Reis: a predominantly fission origin. Nature, **234**:540-543, 1971b.

BIBRING, J.P., BROWNLEE, D.E. and MAURETTE, M. — HVEM and HVES observations of U-2 stratospheric dust grains. Lunar and Planetary Sci. IX, pp. 79-81, 1978.

BINNS, R.A., DAVIS, R.J. and REED, S.J.B. — Ringwoodite, natural $(Mg, Fe)_2SiO_4$ spinel in the Tenham meteorite. Nature, **221**:943-944, 1969.

BIRCK, J.L. and ALLÈGRE, C.J. — Chronology and chemical history of the parent body of basaltic achondrites studied by the $^{87}Rb/^{87}Sr$ method. Earth Planet. Sci. Lett., **39**:37-51, 1978.

BIRCK. J.L., LORIN, J.C. and ALLÈGRE, C. — Potassium isotopic determination in some meteoritic and lunar samples: evidence for irradiation effects. Meteoritics, **12**:179-180, 1977.

BIRCK, J.L., MINSTER, J.F. and ALLÈGRE, C.J. — $^{87}Rb/^{87}Sr$ chronology of achondrites. Meteoritics, **10**:364-365, 1975.

BLANDER, M., PLANNER, H.N., KEIL, K., NELSON, L.S. and RICHARDSON, N.L. — The origin of chondrules: experimental investigation of metastable liquids in the system Mg_2SiO_4–SiO_2. Geochim. Cosmochim. Acta, **40**:889-896, 1976.

BOCTOR, N.Z., MEYER, H.O.A. and KULLERUD, G. — Lafayette meteorite: petrology and opaque mineralogy. Earth Planet. Sci. Lett., **32**:69-76, 1976.

BOECKL, R. — Terrestrial age of nineteen stony meteorites derived from their radiocarbon content. Nature, **236**:25-26, 1972.

BOGARD, D.D. and HUSAIN, L. — A new 1.3 Aeon-young achondrite. Geophys. Res. Lett., **4**:69-71, 1977.

BOSTRÖM, K. and FREDRIKSSON, K. — Surface conditions of the Orgueil meteorite parent body as indicated by mineral associations. Smithsonian Misc. Coll., **151**:1-39, 1966.

BRETT, R. — Cohenite in meteorites: a proposed origin. Science, **153**:60-62, 1966.

BRETT, R. and HIGGINS, G.T. — Cliftonite in meteorites: a proposed origin. Science, **156**:819-820, 1967.

BRETT, R., HUEBNER, J.S. and SATO, M. — Measured oxygen fugacities of the Angra dos Reis achondrite as a function of temperature. Earth Planet. Sci. Lett. **35**:363-368, 1977.

BROWN, H. — The density and mass distribution of meteoritic bodies in the neighborhood of the Earth's orbit. J. Geophys. Res., **66**:1316-1317, 1961.

BROWN, H., KULLERUD, G. and NICHIPORUK, W. — A Bibliography on Meteorites. University of Chicago Press, 1953.

BROWNLEE, D.E., RAJAN, R.S. and TOMANDL, D.A. — A chemical and textural comparison between carbonaceous chondrites and interplanetary dust. *In*: Comets, Asteroids, Meteorites (A.H. Delsemme, Ed.), pp. 137-141, The University of Toledo, Ohio, 1977a.

BROWNLEE, D.E., TOMANDL, D.A. and OLSZEWSKI, E. — Interplanetary dust: a new source of extraterrestrial material for laboratory studies. Proc. Eighth Lunar Sci. Conf., pp. 149-160, 1977b.
BROWNLEE, D.E., HODGE, P.W., BLANCHARD, M.B., BUNCH, T.E. and KYTE, F.T. — Chondritic meteor debris in abyssal sediments. Lunar and Planetary Sci. IX, pp. 126-127, 1978.
BUCHWALD, V.F. — Handbook of Iron Meteorites. Vols. 1-3, University of California Press, Berkeley-Los Angeles-London, 1975.
BUNCH, T.E. and REID, A.M. — The Nakhlites. Part 1: petrography and mineral chemistry. Meteoritics, **10**:303-315, 1975.
BUNCH, T.E., KEIL, K. and SNETSINGER, K.G. — Chromite composition in relation to chemistry and texture of ordinary chondrites. Geochim. Cosmochim. Acta, **31**:1569-1582, 1967.
BURRAGATO, F., CAVARRETTA, G. and FUNICIELLO, R. — The new Brazilian achondrite of Governador Valadares, Minas Gerais. Meteoritics, **10**:374-375, 1975.
BUSECK, P.R. and HOLDSWORTH, E.J. — Mineralogy and petrology of the Yilmia enstatite chondrite. Meteoritics, **7**:429-447, 1972.
BUSECK, P.R. and KEIL, K. — Meteoritic rutile. Amer. Mineral., **51**:1506-1511, 1966.
CABRI, L.J. — A new copper-iron sulfide. Econ. Geol., **62**:910-925, 1967.
CAMERON, A.G.W. — The formation of the sun and planets. Icarus, **1**:13-69, 1962.
CARVER, E.A. — History of Ca-rich achondrites from solid state irradiation damage caused by charged particles. Ph.D. Thesis. University of Chicago, 1974.
CARVER, E.A. and ANDERS, E. — Serra de Magé: a meteorite with an unusual history. Earth Planet. Sci. Lett., **8**:214-220, 1970.
CARVER, E.A. and ANDERS, E. — Nuclear tracks in the Angra dos Reis and Moore County meteorites. Geochim. Cosmochim. Acta, **40**:935-944, 1976a.
CARVER, E.A. and ANDERS, E. — Fission-track ages of four meteorites. Geochim. Cosmochim. Acta, **40**:467-477, 1976b.
CASSIDY, W.A., OLSEN, E. and YANAI, K. — Antarctica: a deep-freeze storehouse for meteorites. Science, **198**:727-731, 1977.
CAVARRETTA, G., FORNASERI, M., FUNICIELLO, R. and TOLOMEO, L. — The chondritic shower of Lajeado Ipiranga, Paraná, Brazil. Meteoritics, **10**:380, 1975.
CEPLECHA, Z. — Multiple fall of Pribram meteorites photographed. Bull. Astron. Inst. Czech., **12**:21-47, 1961.
CHAPMAN, C.R. — The nature of asteroids. Scient. Amer., **232**:24-33, 1975.
CHAPMAN, C.R. — The evolution of asteroids as meteorite parent bodies. *In*: Comets, Asteroids, Meteorites (A.H. Delsemme, Ed.), pp. 265-275, The University of Toledo, Ohio, 1977.
CHAPMAN, C.R., McCORD, T.B. and JOHNSON, T.V. — Asteroid spectral reflectivities. Astron. J., **78**:126-140, 1973a.
CHAPMAN, C.R., McCORD, T.B. and PIETERS, C. — Minor planets and related objects X. Spectrophotometric study of the composition of (1685) Toro. Astron. J., **78**:502-505, 1973b.
CHLADNI, E.F.F. — Über den Ursprung der von Pallas gefundenen und anderer ihr ähnlicher Eisenmassen und über einige damit in Verbindung stehende Naturerscheinungen. J.F. Hartknoch, Riga, 1794.
CHRISTOPHE MICHEL-LEVY, M. and LORIN, J.C. — El Quemado, a new type of stone meteorite fallen near Acapulco. Meteoritics, **13**:411-412, 1979.
CLARKE, R.S., JAROSEWICH, E., MASON, B., NELEN, J., GOMEZ, M. and HYDE, J.R. — The Allende, Mexico, meteorite shower. Smithson. Contr. Earth Sci., **5**:1-53, 1971.
CLAYTON, R.N. — Isotopic anomalies in the early solar system. Ann. Rev. Nucl. Part. Sci., **28**:501-522, 1978.
CLAYTON, R.N. and MAYEDA, T.K. — Genetic relations between the moon and meteorites. Proc. Sixth Lunar Sci. Conf., pp. 1761-1769, 1975.

CLAYTON, R.N., GROSSMAN, L. and MAYEDA, T.K., – A component of primitive nuclear composition in carbonaceous meteorites. Science, **182**:485-488, 1973.
CLAYTON, R.N., ONUMA, N. and MAYEDA, T.K. – A classification of meteorites based on oxygen isotopes. Earth Planet. Sci. Lett., **30**:10-18, 1976.
COHEN, E. – Meteoritenkunde. Vols. I, II and III. Schweizerbart, Stuttgart, 1894, 1903, 1905.
CONSOLMAGNO, G.J. and DRAKE, M.J. – Composition and evolution of the eucrite parent body: evidence from rare earth elements. Geochim. Cosmochim. Acta, **41**:1271-1282, 1977.
COUTINHO, J.M.V. and ARID, F.M. – Um condrito de São José do Rio Preto. Bol. Soc. Bras. Geol., **12**:75-85, 1963.
CRAIG, H. – Petrological and compositional relationships in meteorites. *In*: Isotopic and Cosmic Chemistry (H. Craig, S.L. Miller and G.J. Wasserburg, Eds.), pp. 401-451, North-Holland, Amsterdam, 1964.
DERBY, O.A. – Meteoritos Brasileiros. Revista do Observatório, pp. 1-22, 1888a.
DERBY, O.A. – Brazilian meteorites. Amer. J. Sci., **36**:157, 1888b.
DODD, R.T. – Metamorphism of ordinary chondrites. Geochim. Cosmochim. Acta, **33**:161-203, 1969.
DODD, R.T. – Calc-aluminous insets in olivine of the Sharps chondrite. Mineral. Mag., **38**:451-458, 1971.
DODD, R.T., GROVER, J.E. and BROWN, G.E. – Pyroxenes in the Shaw (L-7) chondrite. Geochim. Cosmochim. Acta, **39**:1585-1594, 1975.
DODD, R.T., VAN SCHMUS, W.R. and MARVIN, U.B. – Merrihueite, a new alkali-ferromagnesian silicate from the Mezö-Madaras chondrite. Science, **149**:972-974, 1965.
DOWTY, E. – Phosphate in Angra dos Reis: structure and composition of the $Ca_3(PO_4)_2$ minerals. Earth Planet. Sci. Lett., **35**:347-351, 1977.
DROZD, R.J. and PODOSEK, F.A. – Primordial ^{129}Xe in meteorites. Earth Planet. Sci. Lett., **31**:15-30, 1976.
DROZD, R.J. and PODOSEK, F.A. – Systematics of iodine-xenon dating. Geochem. J., **11**:231-237, 1977.
DU FRESNE, E.R. and ANDERS, E. – On the chemical evolution of the carbonaceous chondrites. Geochim. Cosmochim. Acta, **26**:1085-1114, 1962.
DUKE, M.B. – Petrology of basaltic achondrites. Ph.D. Thesis. California Institute of Technology, 1963.
DUKE, M.B. and BRETT, R. – Metallic copper in stony meteorites. U.S. Geol. Survey, Prof. Paper No. **525-B**:B101-103, 1965.
DUKE, M.B. and SILVER, L.T. – Petrology of eucrites, howardites and mesosiderites. Geochim. Cosmochim. Acta, **31**:1637-1665, 1967.
DYAKONOVA, M.I. – The chemical composition of seven stony meteorites in the collection of the Committee on Meteorites of the Academy of Sciences. Meteoritika, **25**:129-133, 1964.
EBERHARDT, P., GEISS, J. and GRÖGLER, N. – Further evidence on the origin of trapped gases in the meteorite Khor Temiki. J. Geophys. Res., **70**:4375-4378, 1965.
EHMANN, W.D., BAEDECKER, P.A. and McKOWN, D.M. – Gold and iridium in meteorites and some selected rocks. Geochim. Cosmochim. Acta, **34**:493-507, 1970.
EZER, D. and CAMERON, A.G.W. – The early evolution of the Sun. Icarus, **1**:422-441, 1963.
EZER, D. and CAMERON, A.G.W. – A study of solar evolution. Can. J. Phys., **43**:1497-1517, 1965.
FARRINGTON, O.C. – Meteorites: Their Structure, Composition, and Terrestrial Relations. Self-published, 1915.
FLEISCHER, R.L., PRICE, P.B. and WALKER, R.M. – Spontaneous fission tracks from extinct ^{244}Pu in meteorites and the early history of the solar system. J. Geophys. Res., **70**:2703-2707, 1965.
FLEISCHER, R.L., PRICE, P.B. and WALKER, R.M. – Nuclear Tracks in Solids. University of California Press, Berkeley-Los Angeles-London, 1975.

FLORAN, R.J., PRINZ, M., HLAVA, P.F., KEIL, K., NEHRU, C.E. and HINTHORNE, J.R. — The Chassigny meteorite: a cumulate dunite with hydrous amphibole-bearing melt inclusions. Geochim. Cosmochim. Acta, **42**:1213-1229, 1978.
FODOR, R.V. and KEIL, K. — Implications of poikilitic textures in LL-group chondrites. Meteoritics, **10**:325-339, 1975.
FODOR, R.V. and KEIL, K. — Carbonaceous and non-carbonaceous lithic fragments in the Plainview, Texas, chondrite: origin and history. Geochim. Cosmochim. Acta, **40**:177-189, 1976a.
FODOR, R.V. and KEIL, K. — A komatiite-like lithic fragment with spinifex texture in the Eva meteorite: origin from a supercooled impact-melt of chondritic parentage. Earth Planet. Sci. Lett., **29**:1-6, 1976b.
FODOR, R.V., KEIL, K. and GOMES, C.B. — Studies of Brazilian meteorites IV. Origin of a dark-colored, unequilibrated lithic fragment in the Rio Negro chondrite. Rev. Bras. Geoc., **7**:45-57, 1977.
FODOR, R.V., KEIL, K. and JAROSEWICH, E. — The Oro Grande, New Mexico, chondrite and its lithic inclusion. Meteoritics, **7**:495-507, 1972.
FODOR, R.V., KEIL, K., WILKENING, L.L., BOGARD, D.D. and GIBSON, E.K. — Origin and history of a parent body regolith breccia: Carbonaceous and noncarbonaceous lithic fragments in the Abbot, New Mexico, chondrite. *In*: Tectonics and mineral resources of Southwestern North America. New Mexico Geol. Soc., Spec. Publ., **6**:206-218, 1976.
FOWLER, W.A., GREENSTEIN, J.L. and HOYLE, F. — Nucleosynthesis during the early history of the solar system. Geophys. J. Royal Astron. Soc., **6**:148-220, 1962.
FREDRIKSSON, K. and KEIL, K. — The light-dark structure in the Pantar and Kapoeta stone meteorites. Geochim. Cosmochim. Acta, **27**:717-739, 1963.
FREDRIKSSON, K. and KEIL, K. — The iron, magnesium, calcium and nickel distribution in the Murray carbonaceous chondrite. Meteoritics, **2**:201-217, 1964.
FREDRIKSSON, K. and REID, A.M. — A chondrule in the Chainpur meteorite. Science, **149**:856-860, 1965.
FRONDEL, C. and MARVIN, U.B. — Lonsdaleite, a hexagonal polymorph of diamond. Nature, **214**:587-589, 1967.
FRULAND, R. and WALTON, W. — A possible new silico-phosphate mineral in the Rose City meteorite. Meteoritics, **11**:283-285, 1976.
FUCHS, L.H. — Djerfisherite, alkali copper-iron sulfide, a new mineral from the Kota-Kota and St. Mark's enstatite chondrites. Science, **153**:166-167, 1966.
FUCHS, L.H. — Occurrence of cordierite and aluminous orthoenstatite in the Allende meteorite. Amer. Mineral., **54**:1645-1653, 1969a.
FUCHS, L.H. — X-ray crystallographic evidence for the meteoritic occurrence of nepheline. Meteoritics, **4**:176-177, 1969b.
FUCHS, L.H. — Occurrence of wollastonite, rhönite and andradite in the Allende meteorite. Amer. Mineral., **56**:2053-2068, 1971.
FUCHS, L.H. and BLANDER, M. — The mineralogy of a molybdenite and rhönite-bearing calcium-aluminum rich inclusion (CAI) in the Allende meteorite. Meteoritics, **11**:285-286, 1976.
FUCHS, L.H., FRONDEL, C. and KLEIN, C. — Roedderite, a new mineral from the Indarch meteorite. Amer. Mineral., **51**:949-955, 1966.
FUCHS, L.H., JENSEN, K.J. and OLSEN, E. — Mineralogy and composition of the Murchison meteorite. Meteoritics, **5**:198, 1970.
FUCHS, L.H., OLSEN, E. and JENSEN, K.J. — Mineralogy, mineral-chemistry, and composition of the Murchison (C2) meteorite. Smithsonian Contrib. Earth Sci., **10**:1-39, 1973.
FUSE, K. and ANDERS, E. — Aluminium-26 in meteorites — VI. Achondrites. Geochim. Cosmochim. Acta, **33**:653-670, 1969.
GAFFEY, M.J. — Spectral reflectance characteristics of the meteorite classes. J. Geophys. Res., **81**:905-920, 1976.
GAFFEY, M.J. and HAPKE, T.B. — Asteroid surface materials: mineralogical characterization from reflectance spectra. Space Sci. Rev., **21**:555-628, 1978.

GANAPATHY, R. — Rare gases from Serra de Magé feldspar. Earth Planet. Sci. Lett., **8**:221-222, 1970.

GANAPATHY, R. and ANDERS, E. — Ages of calcium-rich achondrites — II. Howardites, nakhlites and the Angra dos Reis angrite. Geochim. Cosmochim. Acta, **33**:775-787, 1969.

GATTERER, A. and JUNKES, J. — Über den Steinmeteoriten von Rio Negro. Pontificia Acad. Sci., Commentationes, **4**:191-223, 1940.

GOLDSTEIN, J.I. — The formation of the kamacite phase in metallic meteorites. J. Geophys. Res., **70**:6223-6232, 1965.

GOLDSTEIN, J.I. and AXON, H.J. — The Widmanstätten figure in iron meteorites. Naturwiss., **60**:313-321, 1973.

GOLDSTEIN, J.I. and OGILVIE, R.E. — A reevaluation of the iron-rich portion of the Fe-Ni system. Trans. AIME, **223**:2083-2087, 1965a.

GOLDSTEIN, J.I. and OGILVIE, R.E. — The growth of the Widmanstätten pattern in metallic meteorites. Geochim. Cosmochim. Acta, **29**:893-920, 1965b.

GOLDSTEIN, J.I. and SHORT, J.M. — Cooling rates of 27 iron and stony-iron meteorites. Geochim. Cosmochim. Acta, **31**:1001-1023, 1967.

GOMES, C.B. and KEIL, K. — Studies of Brazilian meteorites X. Mineralogy and petrology of the Sete Lagoas, Minas Gerais, chondrite. Bol. IG-USP, **8**:77-82, 1977a.

GOMES, C.B. and KEIL, K. — Estudos de meteoritos brasileiros XI. Revisão de um programa de pesquisa. Ciência e Cultura, **29**:1094-1103, 1977b.

GOMES, C.B. and KEIL, K. — Meteorito Paranaíba, Matro Grosso: uma revisão. An. Acad. brasil. Ciênc., **50**:114, 1978.

GOMES, C.B., KEIL, K. and JAROSEWICH, E. — Studies of Brazilian meteorites VII. Mineralogy, petrology and chemistry of the Uberaba, Minas Gerais, chondrite. An. Acad. brasil. Ciênc., **49**:269-274, 1977a.

GOMES, C.B., KEIL, K., JAROSEWICH, E. and CURVELLO, W.S. — Studies of Brazilian meteorites VIII. Mineralogy, petrology and chemistry of the Itapicuru Mirim, Maranhão, chondrite. An. Acad. Brasil, Ciênc., **49**:407-412, 1977b.

GOMES, C.B., KEIL, K., JAROSEWICH, E. and CURVELLO, W.S. — Studies of Brazilian meteorites IX. Mineralogy, petrology and chemistry of the Macau, Rio Grande do Norte, chondrite. An. Acad. brasil. Ciênc., **49**:575-579, 1977c.

GOMES, C.B., KEIL, K., JAROSEWICH, E. and CURVELLO, W.S. — Studies of Brazilian meteorites VI. Mineralogy, petrology and chemistry of the Patrimônio, Minas Gerais, chondrite. J. Mineral., Vol. Djalma Guimarães, **7**:67-71, 1978a.

GOMES, C.B., KEIL, K., CONRAD, G.H., MELLO, Z.F. and BARRETO, A. — O condrito Parambu, Ceará: uma brecha polimítica originada por impacto. XXX Congr. Bras. Geol., **1**:97-98, 1978b.

GOMES, C.B., KEIL, K., CONRAD, G.H., MELLO, Z.F. and BARRETO, A. — Studies of Brazilian meteorites XIX. The Parambu chondrite, an impact breccia. In preparation.

GOMES, C.B., KEIL, K., RUBERTI, E., JAROSEWICH, E. and SILVA, J.M.L.U. — Studies of Brazilian meteorites XVI. Mineralogy, petrology and chemistry of the Ipiranga, Paraná, chondrite. Chem. Erde, **37**:265-270, 1978c.

GOMES, C.B., ULBRICH, M.N.C., KEIL, K., KIRCHNER, E. and JAROSEWICH, E. — Studies of Brazilian meteorites XV. Mineralogy, petrology and chemistry of the São José do Rio Preto, São Paulo, chondrite. Naturalia, **4**:25-30, 1978d.

GRAHAM, A.L., EASTON, A.J. and HUTCHISON, R. — Forsterite chondrites: the meteorites Kakangari, Mount Morris (Wisconsin), Pontlyfini and Winona. Mineral. Mag., **41**:201-210, 1977a.

GRAHAM, A.L., EASTON, A.J., and HUTCHISON, R. — The Mayo Belwa meteorite: a new enstatite achondrite fall. Mineral. Mag., **41**:487-492, 1977b.

HAGGERTY, S.E. — The Allende meteorite: a new titanate in condensates from the early solar nebula. Meteoritics, **12**:247-248, 1977.

HALLIDAY, I., BLACKWELL, A.T. and GRIFFIN, A.A. — The Innisfree meteorite and the Canadian camera network. J. Royal Astron. Soc. Canada, **72**:15-39, 1978.

HARLOW, G.E., NEHRU, C.E., PRINZ, M., TAYLOR, G.J. and KEIL, K. — Pyroxenes in Serra de Magé: cooling history in comparison with Moama and Moore County. Earth Planet. Sci. Lett., **43**:173-181, 1979.

HAYES, J.M. — Organic constituents of meteorites — a review. Geochim. Cosmochim. Acta, **31**:1395-1440, 1967.
HAZEN, R.M. and FINGER, L.W. — Crystal structure and compositional variation of Angra dos Reis fassaite. Earth Planet. Sci. Lett., **35**:357-362, 1977.
HEIDE, F. — Meteorites. University of Chicago Press, 1964.
HERZOG, G.F. and ANDERS, E. — Absolute scale for radiation ages of stony meteorites. Geochim. Cosmochim. Acta, **35**:605-611, 1971.
HESS, H.H. and HENDERSEN, E.P. — The Moore County meteorite: a further study with comment on its primordial environment. Amer. Mineral., **34**:494-507, 1949.
HEY, M.H. — Catalogue of Meteorites. British Museum (Natural History), 3rd Ed., London, 1966.
HEYMANN, D. and ANDERS, E. — Meteorites with short cosmic-ray exposure ages, as determined from their Al^{26} content. Geochim. Cosmochim. Acta, **31**:1793-1809, 1967.
HEYMANN, D., MAZOR, E. and ANDERS, E. — Ages of calcium-rich achondrites — I. Eucrites. Geochim. Cosmochim. Acta, **32**:1241-1268, 1968.
HIGUCHI, H. and MORGAN, J.W. — Ancient meteoritic component in Apollo 17 boulders. Proc. Sixth Lunar Sci. Conf., pp. 1625-1651, 1975.
HINTENBERGER, J., KÖNIG, H., SCHULTZ, L. and WÄNKE, H. — Radiogene, spallogene und primordiale Edelgase in Steinmeteoriten III. Z. Naturforsch., **21a**:983-989, 1965.
HOHENBERG, C.M. — Xenon from the Angra dos Reis meteorite. Geochim. Cosmochim. Acta, **34**:185-191, 1970.
HOHENBERG, C.M., MUNK, M.N. and REYNOLDS, J.H. — Spallation and fissiogenic xenon and krypton from stepwise heating of the Pasamonte achondrite; the case for extinct plutonium 244 in meteorites; relative ages of chondrites and achondrites. J. Geophys. Res., **72**:3139-3177, 1967.
HOUSEN, K.R., WILKENING, L.L., CHAPMAN, C.R. and GREENBERG, P. — Asteroidal regolith. Icarus. In press.
HUGUENIN, R.L. and McCORD, T.B. — The surfaces of solar system bodies. EOS, **52**:471-483, 1971.
HUNEKE, J.C., PODOSEK, F.A. and WASSERBURG, G.F. — An argon Bouillabaise including ages from the Luna 20 site. Lunar Sci. IV, pp. 403-405, 1973.
HUSS, G. — A recalculation of the frequency of meteorite falls based upon field studies in meteorite collection. Meteoritics, **13**:498, 1978.
HUSSAK, E. — Über einen Meteorsteinfall westlich von Uberaba im Staate Minas. Ann. Naturhis. Hof Museums, Wien, **19**:85-89, 1904.
HUTCHISON, R. — The Angra dos Reis (stone) mineral assemblage and the genesis of stone meteorites. Nature, **240**:58-59, 1972.
HUTCHISON, R. — Angra dos Reis (stone) mineralogy and its bearing on the genesis of chondrites. Meteoritics, **8**:46, 1973.
HUTCHISON, R., BEVAN, A.W.R. and HALL, J.M. — Appendix to the Catalogue of Meteorites. British Museum (Natural History); London, 1977.
JESSBERGER, E.K., HUNEKE, J.C., PODOSEK, F.A. and WASSERBURG, G.J. — High resolution argon analysis of neutron-irradiated Apollo 16 rocks and separated minerals. Proc. Fifth Lunar Sci. Conf., pp. 1419-1449, 1974.
JOHNSON, J.E., SCRYMGOUR, J.M., JAROSEWICH, E. and MASON, B. — Brachina meteorite: a chassignite from South Australia. Rec. S. Austr. Mus., **17**:309-319, 1977.
KEIL, K. — Forschritte in der Meteoritenkunde. Fortschr. Miner., **38**:202-283, 1960.
KEIL, K. — On the phase composition of meteorites. J. Geophys. Res., **67**:4055-4061, 1962a.
KEIL, K. — Quantitative-erzmikroskopische Integrationsanalyse der Chondrite. Chem. Erde, **22**:281-348, 1962b.
KEIL, K. — Mineralogical and chemical relationships among enstatite chondrites. J. Geophys. Res., **73**:6945-6976, 1968.
KEIL, K. — Meteorite Composition. *In*: Handbook of Geochemistry (K.H. Wedepohl, Ed.), Vol. I, pp. 78-115, Springer-Verlag, 1969a.

KEIL, K. — Titanium distribution in enstatite chondrites and achondrites, and its bearing on their origin. Earth Planet. Sci. Lett., **7**:243-248, 1969b.
KEIL, K. and ANDERSEN, C.A. — Electron microprobe study of the Jajh deh Kot Lalu enstatite chondrite. Geochim. Cosmochim. Acta, **29**:621-638, 1965a.
KEIL, K. and ANDERSEN, C.A. — Occurrences of sinoite, Si_2N_2O, in meteorites. Nature, **207**:745, 1965b.
KEIL, K. and BRETT, R. — Heideite, $(Fe,Cr)_{1+x}(Ti, Fe)_2S_4$, a new mineral in the Bustee achondrite. Amer. Mineral., **59**:465-470, 1974.
KEIL, K. and FREDRIKSSON, K. — Electron microprobe analysis of some rare minerals in the Norton County achondrite. Geochim. Cosmochim. Acta, **27**:939-947, 1963.
KEIL, K. and FREDRIKSSON, K. — The iron, magnesium and calcium distribution in coexisting olivines and rhombic pyroxenes of chondrites. J. Geophys. Res., **69**:3487-3515, 1964.
KEIL, K. and FUCHS, L.H. — Hibonite $(Ca_2(Al, Ti)_{24}O_{38})$ from the Leoville and Allende chondritic meteorites. Earth Planet. Sci. Lett., **12**:184-190, 1971.
KEIL, K. and SNETSINGER, K.G. — Niningerite, a new meteoritic sulfide. Science, **155**:451-453, 1967.
KEIL, K., HUSS, G.I. and WIIK, H.B. — The Leoville, Kansas, meteorite: a polymict breccia of carbonaceous chondrites and achondrite. *In*: Meteorite Research (P.M. Millman, Ed.), p. 217, D. Reidel, 1969.
KEIL, K., PRINZ, M. and BUNCH, T.E. — Mineralogy, petrology and chemistry of some Apollo 12 samples. Proc. Second Lunar Sci. Conf., pp. 319-341, 1971.
KEIL, K., KIRCHNER, E., GOMES, C.B. and NELEN, J. — Studies of Brazilian metorites V. Evidence for shock metamorphism in the Paranaíba, Mato Grosso, chondrite. Rev. Bras. Geoc., **7**:256-268, 1977.
KEIL, K., MASON, B., WIIK, H.B. and FREDRIKSSON, K. — The Chainpur meteorite. Amer. Mus. Novitates, **2173**:1-28, 1964.
KEIL, K., KIRCHNER, E., GOMES, C.B., JAROSEWICH, E. and MURTA, R.L.L. — Studies of Brazilian meteorites XIV. Mineralogy, petrology and chemistry of the Conquista, Minas Gerais, chondrite. Meteoritics, **13**:177-187, 1978a.
KEIL, K., LANGE, D., ULBRICH, M.N.C., GOMES, C.B., JAROSEWICH, E., ROISENBERG, A. and SOUZA, M.J. — Studies of Brazilian meteorites XIII. Mineralogy, petrology and chemistry of the Putinga, Rio Grande do Sul, chondrite. Meteoritics, **13**:165-175, 1978b.
KEIL, K., PRINZ, M., PLANNER, H.N., SKAGGS, S. R., DOWTY, E., NELSON, L.S., RICHARDSON, N.L. and BLANDER, M. — A qualitative comparison of textures in lunar chondrules and CO_2-laser-formed synthetic chondrule-like spherules. Spec. Publ. No. 7, Univ. New Mexico, pp. 1-16, 1973.
KEIL, K., PRINZ, M., HLAVA, P.F., GOMES, C.B., CURVELLO, W.S., WASSERBURG, G.J., TERA, F., PAPANASTASSIOU, D.A., HUNEKE, J.C., MURALI, A.V., MA, M.S., SCHMITT, R.A., LUGMAIR, G.W., MARTI, K., SCHEININ, N.B. and CLAYTON, R.N. — Progress by the consorts of Angra dos Reis. Lunar Sci. VII, pp. 443-445, 1976.
KELLY, W.R. and WASSERBURG, G.J. — Evidence for the existence of ^{107}Pd in the early solar system. Geophys. Res. Lett., **5**:1079-1082, 1978.
KERRIDGE, J.F. — Carbonaceous chondrites in the electron microscope. Fifth Intern. Congress on Electron Microscopy. Academic Press, New York, 1962.
KERRIDGE, J.F. — Meteoritic pyrrhotite. Meteoritics, **5**:149-152, 1970.
KING, E.A., CARMAN, M.F. and BUTLER, J.C. — Chondrules in Apollo 14 samples: implications for the origin of chondritic meteorites. Science, **175**:55-56, 1972.
KIRSTEN, T., RIES, D. and FIREMAN, E.L. — Exposure and terrestrial ages of four Allan Hills Antarctic meteorites. Meteoritics, **13**:519-522, 1978.
KÖNIG, H., KEIL, K. and HINTENBERGER, H. — Untersuchungen an Steinmeteoriten mit extrem hohem Edelgasgehalt II. Der Chondrit Tabor. Z. Naturwiss., **17a**:357-358, 1962.
KÖNIG, H., KEIL, K., HINTENBERGER, H., WLOTZKA, F. and BEGEMANN, F. — Untersuchungen an Steinmeteoriten mit extrem hohem Edelgasgehalt I. Der Chondrit Pantar. Z. Naturwiss., **16a**:1124-1130, 1961.

KRINOV, E.L. — Principles of Meteoritics. Pergamon Press, New York, 1960.
KURAT, G. — Einige Chondren aus dem Meteoriten von Mezö-Madaras Geochim. Cosmochim. Acta, **31**:1843-1858, 1967.
KURAT, G., KEIL, K. and PRINZ, M. — Rock 14318: a polymict lunar breccia with chondritic texture. Geochim. Cosmochim. Acta, **38**:1133-1146, 1974.
KURAT, G., KEIL, K., PRINZ, M. and NEHRU, C.E. — Chondrules of lunar origin. Proc. Third Lunar Sci. Conf., pp. 707-721, 1972.
KURODA, P.K. — Nuclear fission in the early history of the earth. Nature, **187**:36-38, 1960.
KVASHA, L.G. — Investigation of the stony meteorite Staroye Boriskino. Meteoritika, **4**:83-96, 1948 (in Russian).
LANGE, D.E., GOMES, C.B. and KEIL, K. — Studies of Brazilian meteorites XX. The brecciated L-group chondrite Mafra revisited. In preparation.
LARIMER, J.W. and RAMBALDI, E.R. — The Shaw chondrite — Some clues to the core-forming process. Meteoritics, **11**:317, 1976.
LATHAM, G.V., EWING, M., PRESS, F., SUTTON, G., DORMAN, J., NAKAMURA, Y., TOKSOZ, N., LAMMLEIN, D. and DUENNEBIER, F. — Passive seismic experiment. *In*: Apollo 16 Preliminary Science Repört, NASA SP-315, 9-1 — 9-29, 1972.
LAUL, J.C., KEAYS, R.R., GANAPATHY, R., ANDERS, E. and MORGAN, J.W. — Chemical fractionations in meteorites — V. Volatile and siderophile elements in achondrites and ocean ridge basalts. Geochim. Cosmochim. Acta, **36**:329-345, 1972.
LEE, T., PAPANASTASSIOU, D.A. and WASSERBURG, G.J. — Demonstration of ^{26}Mg excess in Allende and evidence for ^{26}Ál. Geophys. Res. Lett., **3**:109-112, 1976.
LEE, T., PAPANASTASSIOU, D.A. and WASSERBURG, G.J. — ^{26}Al in the early solar system: fossil or fuel?. Astrophys. J., **211**:L107-L110, 1977.
LEITCH, C.A. and GROSSMAN, L. — Lithic fragments in Supuhee. Meteoritics, **11**:319-320, 1976.
LEONARD, F.C. — A classificational catalog of the meteorite falls of the world. Univ. Calif. Publ. Astron., **2**:1-80, 1956.
LEVIN, B.J., SIMONENKO, A.N. and ANDERS, E. — Farmington meteorite: a fragment of an Apollo asteroid? Icarus, **28**:307-324, 1976.
LEVI-DONATI, G.R. and SIGHINOLFI, G.P. — The meteoritic shower of Parambu, Ceará State, Brazil: mineralogy and petrology. Meteoritics, **9**:1-9, 1974.
LEVI-DONATI, G.R., SHIMA, M. and SIGHINOLFI, G.P. — Brazilian meteorites: the Mafra, Santa Catarina State, chondrite. Meteoritics, **11**:29-41, 1976.
LEWIS, R.S. and ANDERS, E. — Condensation time of the solar nebula from extinct ^{129}I in primitive meteorites. Proc. Nat. Acad. Sci., **72**:268-273, 1975.
LIEBERMANN, K.W. and EHMANN. W.D. — Determination of bromine in stony meteorites by neutron activation. J. Geophys. Res., **72**:6279-6287, 1967.
LIPSCHUTZ, M.E. — Diamonds in the Dyalpur meteorite. Science, **138**:1266-1267, 1962.
LOVERING, J.F. — The Moama eucrite: a pyroxene-plagioclase adcumulate. Meteoritics, **10**:101-114, 1975.
LUDWIG, E. and TSCHERMAK, G. — Der Meteorit von Angra dos Reis. Mineral. Petrog. Mitt., **8**:341-355, 1887.
LUDWIG, E. and TSCHERMAK, G. — Nachtrag zu der Mitteilung über den Meteoriten von Angra dos Reis. Mineral. Petrog. Mitt., **28**:110-114, 1909.
LUGMAIR, G.W. and MARTI, K. — Sm-Nd-Pu timepieces in the Angra dos Reis meteorite. Earth Planet. Sci. Lett., **35**:273-284, 1977.
LUGMAIR, G.W., SCHEININ, N.B. and CARLSON, R.W. — Sm-Nd systematics of the Serra de Magé eucrite. Meteoritics, **12**:300-301, 1977.
MA, M.-S. and SCHMITT, R.A. — Genesis of the cumulate eucrites Serra de Magé and Moore County: a geochemical study. Meteoritics, **14**:81-89, 1979.
MA, M.-S., MURALI, A.V. and SCHMITT, R.A. — Genesis of the Angra dos Reis and other achondritic meteorites. Earth Planet. Sci. Lett., **35**:331-346, 1977.
MANHES, G., TATSUMOTO, M., UNRUH, D., BIRCK, J.L. and ALLÈGRE, C.J.

— Comparative ^{238}U-^{206}Pb, ^{235}U-^{207}Pb, ^{232}Th-^{208}Pb, ^{206}Pb-^{207}Pb, and ^{87}Rb-^{87}Sr ages of basaltic achondrites and early evolution of the solar system. Lunar Sci. VI, 546-547, 1975.
MARVIN, U.B., WOOD, J.A. and DICKEY, J.S. — Ca-Al-rich phases in the Allende meteorite. Earth Planet. Sci. Lett., **7**:346-350, 1970.
MASON, B. — Meteorites. John Wiley & Sons, New York, 1962.
MASON, B. — Olivine composition in chondrites. Geochim. Cosmochim. Acta, **27**:1011-1023, 1963.
MASON, B. — Feldspar in chondrites. Science, **148**:943, 1965.
MASON, B. — The enstatite chondrites. Geochim. Cosmochim. Acta, **30**:23-39, 1966.
MASON, B. — Meteorites. Amer. Sci., **55**:429-455, 1967a.
MASON, B. — Extraterrestrial mineralogy. Amer. Mineral., **52:**307-325, 1967b.
MASON, B. — Olivine composition in chondrites — a supplement. Geochim. Cosmochim. Acta, **31**:1100-1103, 1967c.
MASON, B. — Handbook of Elemental Abundances in Meteorites. Gordon and Breach, New York, 1971.
MASON, B. — The mineralogy of meteorites. Meteoritics, **7**:309-326, 1972.
MASON, B. — Achondrite (unique). *In:* Antartic Meteorite Newsletter (Antarctic Meteorite Working Group), **1**, No. **2**:9, NASA Johnson Space Center, Houston, Texas, 1978.
MAZOR, E., HEYMANN, D. and ANDERS, E. — Noble gases in carbonaceous chondrites. Geochim. Cosmochim. Acta, **34**:781-824, 1970.
McCARTHY, T.S., ERLANK, A.J. and WILLIS, J.P. — On the origin of eucrites and diogenites. Earth Planet. Sci. Lett., **18**:433-442, 1973.
McCARTHY, T.B., ERLANK, A.J., WILLIS, J.P. and AHRENS, L.H. — New chemical analyses of six achondrites and one chondrite. Meteoritics, **9**:215-221, 1974.
McCORD, T.B. and GAFFEY, M.J. — Asteroids: surface composition from reflection spectroscopy. Science, **186**:352-355, 1974.
McCORD, T.B., ADAMS, J.B. and JOHNSON, T.V. — Asteroid Vesta: spectral reflectivity and compositional implications. Science, **168**:1445-1447, 1970.
McCROSKY, R.E. — The distribution of large meteoritic bodies. Smiths. Astrophys. Observ. Spec. Rep. 280, Cambridge, Mass., 1968.
McCROSKY, R.E., POSEN, A., SCHWARTZ, G. and SHAO, G.-Y. — Lost City meteorite — its recovery and a comparison with other fireballs. J. Geophys. Res., **76**:4090-4108, 1971.
MELFI, A.J. — Condrito caído no Município de Mafra, SC. Bol. Soc. Bras. Geol., **14**:41-52, 1965.
MENEZES, V. — A probable meteorite fall in Brazil. Sky and Telescope, **17**:10, 1957.
MILLMAN, P.M. — The chemical composition of cometary meteorites. *In*: Comets, Asteroids, Meteorites (A.H. Delsemme, Ed.), pp. 127-132, The University of Toledo, Ohio, 1977.
MOORE, C.B. and LEWIS, C.F. — Carbon abundances in chondritic meteorites. Science, **149**:317-318, 1965.
MOORE, C.B. and LEWIS, C.F. — The distribution of total carbon content in enstatite chondrites. Earth Planet. Sci. Lett., **1**:376-378, 1966.
MORAES, L.J. and GUIMARÃES, D. — Meteorito caído na Serra de Magé, Município de Pesqueira, Pernambuco. Bol. Inst. Bras. Sci., **2**:356-360, 1926.
MORGAN, J.W. and LOVERING, J.F. — Uranium and thorium abundances in stony meteorites 2. The achondritic meteorites. J. Geophys. Res., **69**:1989-1994, 1964.
MORGAN, J.W. and LOVERING, J.F. — Uranium and thorium in achondrites. Geochim. Cosmochim. Acta, **37**:1697-1707, 1973.
MORGAN, J.W., GANAPATHY, R., HIGUCHI, H. and KRÄHENBÜHL, U. — Volatile and siderophile trace elements in anorthositic rocks from Fiskenaesset, West Greenland: comparison with lunar and meteoritic analogues. Geochim. Cosmochim. Acta, **40**:861-887, 1976.
MORGAN, J.W., HIGUCHI, H., TAKAHASHI, H. and HERTOGEN, J. — A "chondritic" eucrite parent body: inference from trace elements. Geochim. Cosmochim. Acta, **42**:27-38, 1978.

MÜLLER, H.W. and ZÄHRINGER, J. — Rare gases in stony meteorites. *In*: Meteorite Research (P.M. Millman, Ed.), pp. 845-856, D. Reidel, 1969.
MÜLLER, W.F., KURAT, G. and KRACHER, A. — Crystal structure and composition of cronstedtite from the Cochabamba carbonaceous chondrite. Meteoritics, **12**:322, 1977.
MUNK, M.N. — Argon, krypton and xenon in Angra dos Reis, Nuevo Laredo and Norton Country achondrites: the case for two types of fission xenon in achondrites. Earth Planet. Sci. Lett., **3**:457-465, 1967.
MURTA, R.L.L. — O meteorito de Conquista. XXV Congr. Bras. Geol., Resumo das Comunicações, Bol. Esp. **1**:122, 1971.
NAGATA, T. — Yamato meteorites collected in Antarctica in 1969. Mem. Natl. Institute of Polar Research, Tokyo, Japan, Spec. Iss., No. 5, 1975.
NAGATA, T. — Proceedings of the Second Symposium on the Yamato Meteorites. Mem. Natl. Institute of Polar Research, Tokyo, Japan, Spec. Iss., No. 8, 1978.
NAGY, B. — Carbonaceous Meteorites. Elsevier Publ. Comp., 1975.
NAGY, B. and ANDERSEN, C.A. — Electron probe microanalysis of some carbonate, sulfate, and phosphate minerals in the Orgueil meteorite. Amer. Mineral., **49**:1730-1736, 1964.
NAGY, B., FREDRIKSSON, K., UREY, H.C., CLAUS, G., ANDERSEN, C.A. and PERCY, J. — Electron probe microanalysis of organized elements in the Orgueil meteorite. Nature, **198**:121-125, 1963.
NELSON, L.S., BLANDER, M., SKAGGS, S.R. and KEIL, K. — Use of a CO_2 laser to prepare chondrule-like spherules from supercooled molten oxide and silicate droplets. Earth Planet. Sci. Lett., **14**:338-344, 1972.
NOONAN, A.F. — Glass particles and shock features in the Bununu howardite. Meteoritics, **9**:233-242, 1974.
NOONAN, A.F., RAJAN, S. and FREDRIKSSON, K. — Agglutinates in ordinary chondrites. Meteoritics, **13**:587, 1978.
NORDEMANN, D. and SAKAI, T. — A radioactividade do meteorito Marília (queda do dia 5 de outubro de 1971) medida por espectrometria gama. Rev. Bras. Geoc., **3**:48-52, 1973.
OLIVEIRA, E. — Collecções de meteoritos do Museu Nacional, do Serviço Geológico do Brasil e da Escola de Minas. An. Acad. brasil. Ciênc., **3**:33-56, 1931.
OLSEN, E., HUEBNER, J.S., DOUGLAS, J.A.V. and PLANT, A.G. — Meteoritic amphiboles. Amer. Mineral., **58**:869-872, 1973.
OLSEN, E., NOONAN, A., FREDRIKSSON, K., JAROSEWICH, E. and MORELAND, G. — Eleven new meteorites from Antarctica, 1976-1977. Meteoritics, **13**:209-225, 1978.
OSTIC, R.G. — Physical conditions in gaseous spheres. Monthly Notices Roy. Astron. Soc., **131**:191-197, 1965.
PAAR, W., KEIL, K., GOMES, C.B. and JAROSEWICH, E. — Studies of Brazilian meteorites II. The Avanhandava chondrite: mineralogy, petrology and chemistry. Rev. Bras. Geoc., **6**:201-210, 1976.
PAPANASTASSIOU, D.A. — The determination of small time differences in the formation of planetary objects. Ph. D. Thesis, California Institute of Technology, 1970.
PAPANASTASSIOU, D.A. and WASSERBURG, G.J. — Initial Sr isotopic abundances and the resolution of small differences in the formation of planetary objects. Earth Planet. Sci. Lett., **5**:361-376, 1969.
PELLAS, P., POUPEAU, G., LORIN, G.C., REEVES, H. and AUDOUZE, J. — Primitive low-energy particle irradiation of meteoritic crystals. Nature, **223**:272-274, 1969.
PERRY, S.H. — The metallography of meteoritic iron. Bull. U.S. National Museum, **184**:1-206, 1944.
PODOSEK, F.A. — Dating of meteorites by the high-temperature release of iodine-correlated Xe^{129} Geochim. Cosmochim. Acta, **34**:341-365, 1970.
PODOSEK, F.A. — Gas retention chronology of Petersburg and other meteorites. Geochim. Cosmochim. Acta, **36**:755-772, 1972.
PODOSEK, F.A. — Thermal history of the Nakhlites by the $^{40}Ar/^{39}Ar$ method. Earth Planet. Sci. Lett., **19**:135-144, 1973.

PODOSEK, F.A. and HUNEKE, J.C. — Isotopic composition of ^{244}Pu fission xenon in meteorites: reevaluation using lunar spallation xenon systematics. Earth Planet. Sci. Lett., **12**:73-82, 1971.
PODOSEK, F.A. and LEWIS, R.S. — ^{129}I and ^{244}Pu abundances in white inclusions of the Allende meteorite. Earth Planet. Sci. Lett., **15**:101-109, 1972.
POKRZYWNICKI, J. — The specific gravity of meteorites. Acta Geophys. Polonica, **6**:127-152, 1959.
PRINZ, M., FODOR, R.V. and KEIL, K. — Comparison of lunar rocks and meteorites: implications to histories of the moon and parent meteorite bodies. Soviet-American Conf. Cosmochem. of the Moon and Planets, NASA SP-370, **1**:183-199, 1977a.
PRINZ, M., KEIL, K., HLAVA, P.F., BERKLEY, J.L., GOMES, C.B. and CURVELLO, W.S. — Studies of Brazilian meteorites III. Origin and history of the Angra dos Reis achondrite. Earth Planet. Sci. Lett., **35**:317-330, 1977b.
PRINZ, M., NEHRU, C.E., BERKLEY, J.L., KEIL, K., JAROSEWICH, E. and GOMES, C.B. — Petrogenesis of the Serra de Magé cumulate eucrite. Meteoritics, **12**:341, 1977c.
PRINZ, M., NEHRU, C.E., BERKLEY, J.L., KEIL, K., JAROSEWICH, E. and GOMES, C.B. — Studies of Brazilian meteorites XVIII. Petrogenesis of the Serra de Magé cumulate eucrite. In preparation.
PRIOR, G.T. — The meteoric stones of El Nakhla El Baharia, Egypt. Mineral. Mag., **16**:274-281, 1912.
RAJAN, R.S. — On the irradiation history and origin of gas-rich meteorites. Geochim. Cosmochim. Acta, **38**:777-788, 1974.
RAMBALDI, E.R. — Trace element content of metals from H- and LL- group chondrites. Earth Planet. Sci. Lett., **36**:347-358, 1977.
RAMBALDI, E.R. and CENDALES, M. — Tungsten in ordinary chondrites. Earth Planet. Sci. Lett., **36**:372-380, 1977.
RAMDOHR, P. — The opaque minerals in stony meteorites. J. Geophys. Res., **68**:2011-2036, 1963a.
RAMDOHR, P. — Beobachtungen am Opakerzbestand einiger Meteoriten besonders von New South Wales. Chem. Erde, **23**:119-145, 1963b.
RAMDOHR, P. — The opaque minerals in stony meteorites. Akademie Verlag, Berlin, 1973.
REED, S.J.B. — Perryite in the Kota-Kota and South Oman enstatite chondrites. Mineral. Mag., **36**:850-854, 1968.
REEVES, H. and AUDOUZE, J. — Early heat generation in meteorites. Earth Planet. Sci. Lett., **4**:135-141, 1968.
REYNOLDS, J.H. — Determination of the age of the elements. Phys. Rev. Lett., **4**:8-10, 1960a.
REYNOLDS, J.H. — I-Xe dating of meteorites. J. Geophys. Res., **65**:3843-3846, 1960b.
REYNOLDS, J.H. — Xenology. J. Geophys. Res., **68**:2939-2956, 1963.
RICHARDSON, S.M. — Vein formation in the C1 carbonaceous chondrites. Meteoritics, **13**:141-159, 1978.
RINGWOOD, A.E. — The Novo-Urei meteorite. Geochim. Cosmochim. Acta, **20**:1-4, 1960.
RINGWOOD, A.E. — Genesis of chondritic meteorites. Rev. Geophys., **4**:113-175, 1966.
RINGWOOD, A.E. — Composition and origin of the earth. Res. School of Earth Sci., Australian Natl. University, Canberra, Publ. 1299, 1977.
ROISENBERG, A. — Meteorito de Putinga, R.G. do Sul. XXIV Congr. Bras. Geol., Resumo Comunicações, Bol. **1**:327-329, 1970.
ROSS, J.E. and ALLER, L.H. — The chemical composition of the sun. Science, **191**:1223-1229, 1976.
SANTOS, R.R. — Notícia sobre o meteorito condrítico de Conceição Aparecida, MG. XXVI Congr. Bras. Geol., Resumo das Comunicações, p. 26, 1972.
SCHMIDT, R.A. — A survey of data on microscopic extraterrestrial particles. NASA TN D-2719, 132 p., 1965.

SCHMIDT, R.A. and KEIL, K. — Electron microprobe study of spherules from Atlantic Ocean sediments. Geochim. Cosmochim. Acta, **30**:471-478, 1966.
SCHMITT, R.A., GOLES, G.G., SMITH, R.H. and OSBORN, T.W. — Elemental abundances in stone meteorites. Meteoritics, **7**:131-213, 1972.
SCHNETZLER, C.C. and PHILPOTTS, J.A. — Genesis of the calcium-rich achondrites in light of rare-earth and barium concentrations. *In*: Meteorite Research (P.M. Millman, Ed.), pp. 206-216, D. Reidel, 1969.
SCHRAMM, D.N., TERA, F. and WASSERBURG, G.J. — The isotopic abundance of ^{26}Mg and limits on the ^{26}Al in the early solar system. Earth Planet. Sci. Lett., **10**:44-59, 1970.
SHIMA, M., JOCHUM, K.P., SIGHINOLFI, G.P. and HINTENBERGER, H. — The chemical composition of major elements and heavy trace metals in chondrites Parambu and Marília. Meteoritics, **9**:199-207, 1974.
SIGNER, P. and SUESS, H.E. — Rare gases in the sun, in the atmosphere, and in meteorites. *In*: Earth Science and Meteoritics (J. Geiss and E.D. Goldberg, Eds.), pp. 241-272, North-Holland, Amsterdam, 1963.
SMALES, A.A., MAPPER, D., WEBB, M.S.W., WEBSTER, R.K. and WILSON, J.D. — Elemental composition of lunar surface material. Proc. Apollo 11 Lunar Sci. Conf., pp. 1575-1581, 1970.
SMITH, J.V. and MASON, B. — Pyroxene-garnet transformation in Coorara meteorite. Science, **168**:832-833, 1970.
SNETSINGER, K.G. and KEIL, K. — Ilmenite in ordinary chondrites. Amer. Mineral., **54**:780-786, 1969.
SNETSINGER, K.G., KEIL, K. and BUNCH, T.E. — Chromite from "equilibrated" chondrites. Amer. Mineral., **52**:1322-1331, 1967.
SONETT, C.P., COLBURN, D.S., SCHWARTZ, K. and KEIL, K. — The melting of asteroidal-sized bodies by unipolar dynamo induction from a primordial T-Tauri sun. Astrophys. Space Sci., **7**:446-448, 1970.
SRINIVASAN, B. and ANDERS, E. — Noble gases in the Murchison meteorite: possible relics of s-process nucleosynthesis. Science, **201**:51-56, 1978.
STEELE, I.M. and SMITH, J.V. — Ibitira eucrite: mineralogy and comparison with lunar samples and other eucrites. Lunar Sci. VII, pp. 833-835, 1976a.
STEELE, I.M. and SMITH, J.V. — Mineralogy of the Ibitira eucrite and comparison with other eucrites and lunar samples. Earth Planet. Sci. Lett., **33**:67-78, 1976b.
STOLPER, E. — Experimental petrology of eucritic meteorites. Geochim. Cosmochim. Acta, **41**:587-611, 1977.
STÖRZER, D. and PELLAS, P. — Cosmic ray and fission tracks in Angra dos Reis. Meteoritics, **11**:185-186, 1976.
STÖRZER, D. and PELLAS, P. — Angra dos Reis: plutonium distribution and cooling history. Earth Planet. Sci. Lett., **35**:285-293, 1977.
SUESS, H.E. and WÄNKE, H. — Radiocarbon content and terrestrial age of twelve stony meteorites and one iron meteorite. Geochim. Cosmochim. Acta, **26**:475-480, 1962.
SUESS, H.E., WÄNKE, H. and WLOTZKA, F. — On the origin of gas-rich meteorites. Geochim. Cosmochim. Acta, **28**:595-607, 1964.
SYMES, R.F. and HUTCHISON, R. — Medanitos and Putinga, two South American meteorites. Mineral. Mag., **37**:721-723, 1970.
SZTROKAY, K.I., TOLNAY, U. and FÖLDVARI-VOGL, M. — Mineralogical and chemical properties of the carbonaceous chondrite from Kaba, Hungary. Acta Geol. Acad. Sci. Hungary, **7**:57-103, 1961.
TATSUMOTO, M., KNIGHT, R.J. and ALLÈGRE, C.J. — Time differences in the formation of meteorites as determined from the ratio lead-207 to lead-206. Science, **180**:1279-1283, 1973.
TAYLOR, G.J., KEIL, K., BERKLEY, J.L., LANGE, D.E., FODOR, R.V. and FRULAND, R.M. — The Shaw meteorite: history of a chondrite consisting of impact-melted and metamorphic lithologies. Geochim. Cosmochim. Acta, **43**:323-337, 1979.
TAYLOR, H.P., JR., DUKE, M.B., SILVER, L.T. and EPSTEIN, S. — Oxygen isotope studies of minerals in stony meteorites. Geochim. Cosmochim. Acta, **29**:489-512, 1965.
TAYLOR, S.R. — Lunar Science: A Post-Appollo View. Pergamon Press, 1975.

TERA, F., EUGSTER, O., BURNETT, D.S. and WASSERBURG, G.F. – Comparative study of Li, Na, K, Rb, Cs, Ca, Sr and Ba abundances in achondrites and Apollo 11 lunar samples. Proc. Apollo 11 Lunar Sci. Conf., pp.1637-1657, 1970.

TSCHERMAK, G. – Beiläufige Angabe der Fallzeit der Meteoriten von Angra. Mineral. Petrog. Mitt., **9**:423, 1888.

UREY, H.C. and CRAIG, H. – The composition of the stone meteorites and the origin of meteorites. Geochim. Cosmochim. Acta, **4**:36-82, 1953.

UREY, H.C., MELE, A. and MAYEDA, T. – Diamonds in stone meteorites. Geochim. Cosmochim. Acta, **13**:1-4, 1957.

VAN SCHMUS, W.R. – Mineralogy and petrology of chondritic meteorites. Earth Sci. Rev., **5**:145-184, 1969.

VAN SCHMUS, W.R. and RIBBE, P.H. – The composition and structural state of feldspar from chondritic meteorites. Geochim. Cosmochim. Acta, **32**:1327-1342, 1968.

VAN SCHMUS, W.R. and WOOD, J.A. – A chemical-petrologic classification for the chondritic meteorites. Geochim. Cosmochim. Acta, **31**:747-765, 1967.

VALARELLI, J.V. and ARRUDA, M.R. – Meteorito de Mafra, Santa Catarina. Ciência e Cultura, **17**:132-133, 1965.

VDOVYKIN, G.P. – New hexagonal modification of carbon in meteorites. Geochem. Intern., **6**:915-918, 1969.

VDOVYKIN, G.P. – Ureilites. Space Sci. Rev., **10**:483-510, 1970.

VEDDER, J.F. – Minor objects in the solar system. Space Sci. Rev., **6**:365-414, 1966.

VIDAL, N. – Meteoritos Brasileiros. Bol. Mus. Nac., **12**:91-109, 1936.

VILCSEK, E. and WÄNKE, H. – Cosmic ray exposure ages and terrestrial ages of stone and iron meteorites derived from ^{36}Cl and ^{39}Ar measurements. *In*: Radioactive Dating, pp. 381-393, IAEA, Vienna, 1963.

VOSHAGE, H. – Bestrahlungsalter und Herkunft der Eisenmeteorite. Z. Naturforsch., **22a**:477-506, 1967.

WAI, C.M. – The metal phase of Horse Creek, Mount Egerton, and Norton County enstatite meteorites. Mineral. Mag., **37**:905-908, 1970.

WÄNKE, H., PALME, H., BADDENHAUSEN, H., DREIBUS, G., JAGOUTZ, E., KRUSE, H., SPETTEL, F., TESCHKE, F. and THACKER, R. – Chemistry of Apollo 16 and 17 samples: bulk composition, late stage accumulation and early differentiation of the moon. Proc. Fifth Lunar Sci. Conf., pp. 1307-1335, 1974.

WÄNKE, H., BADDENHAUSEN, H., BALACESCU, A., TESCHKE, F., SPETTEL, B., DREIBUS, G., PALME, H., QUIJANO-RICO, M., KRUSE, H., WLOTZKA, F. and BEGEMANN, F.-F. – Multielement analyses of lunar samples and some implications of the results. Proc. Third Lunar Sci. Conf., pp. 1251-1268, 1972.

WARK, D.A. and LOVERING, J.F. – Refractory/platinum metal grains in Allende calcium-alùminium-rich clasts (CARC's): possible exotic presolar material? Lunar Sci. VII, pp. 912-914, 1976.

WASSERBURG, G.J., PAPANASTASSIOU, D.A. and SANZ, H.G. – Initial strontium for a chondrite and the determination of a metamorphism or formation interval. Earth Planet. Sci. Lett., **7**:33-43, 1969.

WASSERBURG, G.J., TERA, F., PAPANASTASSIOU, D.A. and HUNEKE, J.C. – Isotopic and chemical investigations on Angra dos Reis. Earth Planet. Sci. Lett., **35**:294-316, 1977.

WASSON, J.T. – Formation of ordinary chondrites. Rev. Geophys. Space Phys., **10**:711-759, 1972.

WASSON, J.T. – Meteorites. Springer-Verlag, 1974.

WETHERILL, G.W. – Collisions in the asteroid belt. J. Geophys. Res., **72**:2429-2444, 1967.

WETHERILL, G.W. – Stone meteorites: time of fall and origin. Science, **159:**79-82, 1968a.

WETHERILL, G.W. – Dynamical study of asteroidal and cometary orbits and their relation to the origin of meteorites. *In*: Origin and Distribution of the Elements (L.H. Ahrens, Ed.), pp. 423-443, Pergamon Press, 1968b.

WETHERILL, G.W. — Relationships between orbits and sources of chondritic meteorites. *In*: Meteorite Research (P.M. Millman, Ed.), pp. 573-589, D. Reidel, 1969.
WETHERILL, G.W. — Solar system sources of meteorites and large meteoroids. Ann. Rev. Earth. Planet. Sci., **2**:303-331, 1974.
WETHERILL, G.W. — Fragmentation of asteroids and delivery of fragments to Earth. *In*: Comets, Asteroids, Meteorites (A.H. Delsemme, Ed.), pp. 283-291, The University of Toledo, Ohio, 1977.
WETHERILL, G.W. — Dynamical evidence regarding the relationship between asteroids and meteorites. *In*: Asteroids: An Exploration Assessmet, NASA, Spec. Publ., 1978.
WETHERILL, G.W. and WILLIAMS, J.G. — Evaluation of the Apollo asteroids as sources of stone meteorites. J. Geophys. Res., **73**:635-648, 1968.
WIIK, H.B. — The chemical composition of some stony meteorites. Geochim. Cosmochim. Acta, **9**:279-289, 1956.
WIIK, H.B. — On regular discontinuities in the composition of meteorites. Commun. Phys. Math., **34**:135-145, 1969.
WILKENING, L.L. — Tysnes Island: an unusual clast composed of solidified immiscible, Fe-FeS and silicate melts. Meteoritics, **13**:1-9, 1978.
WILKENING, L.L. and ANDERS, E. — Some studies of an unusual eucrite: Ibitira. Geochim. Cosmochim. Acta, **39**:1205-1210, 1975.
WILKENING, L.L., LAL, D. and REID, A.M. — The evolution of the Kapoeta howardite based on fossil track studies. Earth Planet. Sci. Lett., **10**:334-340, 1971.
WLOTZKA, F. — Phosphates in the Allende carbonaceous chondrite. Meteoritics, **11**:393-394, 1976.
WOOD, J.A. — Physics and Chemistry of Meteorites. *In*: The Solar System — IV. The Moon, Meteorites and Comets (B.M. Middlehurst and G.P. Kuiper, Eds.), pp. 337-401, University of Chicago Press, 1963.
WOOD, J.A. — The cooling rates and parent planets of several iron meteorites. Icarus, **3**:429-459, 1964.
WOOD, J.A. — Chondrites: their metallic minerals, thermal histories, and parent planets. Icarus, **6**:1-49, 1967.
WOOD, J.A. — Meteorites and the Origin of Planets. McGraw-Hill, New York, 1968.
WOODEN, J.L., NYQUIST, L.E., BOGARD, D.D., BANSAL, B.M., WIESMANN, H., SHIH, C.Y., and McKAY, G.A. — Radiometric ages for three achondrites, Chervony Kut, Governador Valadares, and Allen Hills 77005. Lunar and Planetay Sci. X, pp. 1379-1381, 1979.

10. METEORITE INDEX

A Note on the Authors

Klaus Keil directs the Institute of Meteoritics at the University of New Mexico. C. B. Gomes is on the faculty of the Institute of Geosciences at the University of São Paulo. Since 1974 they have studied Brazilian stone meteorites extensively and published some twenty papers in international scientific journals. Dr. Keil has been to the University of São Paulo about once a year since 1974, and his Brazilian co-worker, Dr. Gomes, has been at the University of New Mexico three times for six months each visit.